THE EDGE OF
THE FUTURE

THE EDGE OF THE FUTURE

Popular Science Writing at the Rise of the Twentieth Century

Edited by Larry D. Clark

Articles by Cleveland Moffett, Henry J. W. Dam,
Ida M. Tarbell, Ray Stannard Baker, S. P. Langley,
Sir Robert Ball, Henry Drummond, E. J. Edwards

IRON OWL
BOOKS

Introduction and references:
Copyright ©2014 by Larry D. Clark

Published by Iron Owl Books
Pullman, Wa.
books.iron-owl.com

IRON OWL
BOOKS

ISBN: Paperback 978-0-9912020-0-3 | E-book 978-0-9912020-1-0

Contents

Introduction

By Larry D. Clark

January 2014

ALEXANDER GRAHAM BELL, when asked how his work with electric communication might progress in the future, predicted that thoughts might be transferred from person to person. Less outrageously, he also imagined hearing aids for the deaf and the principles of television signals. The great inventor speculated about the future in 1893 on the pages of *McClure's Magazine*, where readers were given stories on the latest scientific discoveries, inventions, and investigations along with fiction, current issues, and other articles.

In *McClure's*, famed chemist Marcellin Berthelot imagined what synthetic foods would be like in the year 2000 (including steak from a tablet and sugar substitutes). Thomas Edison proposed more efficient extraction of energy from coal than burning. Readers were introduced to automobiles—most of them electric in the America of the late 1890s—and flying machines, the X-ray, the wireless telegraph, and the possibility of life on other planets. They learned about earthquakes and volcanoes and the hunt for the temperature absolute zero. They were taken into volatile nitroglycerin hills to find how dynamite was made under the ever-watchful eye of a man concentrating on a thermometer.

These new inventions, discoveries, and theories impacted the everyday lives and captured the imagination of Americans in the 1890s. The curious and increasingly literate populace were reading more magazines, and *McClure's Magazine*, a new player on the publication scene of the time, delivered to them a new form of science writing.

Started in 1893 by Samuel S. McClure, the magazine rose to prominence quickly on the strength of its editorial content, illustrations, and low price. McClure had an uncanny knack for cultivating writers, bringing authors such as Arthur Conan Doyle, Rudyard Kipling, Willa Cather, Robert Louis Stevenson, and Jack London to American readers. McClure also had a strong interest in communicating science and technical innovations to the masses. McClure (1914) wrote in his autobiography,

> "We made, I think, a more serious effort in the direction of popular science articles than had been made by any magazine before us. *McClure's* was the first popular journal to announce Marconi's discovery of wireless telegraphy, and when that article appeared it was generally regarded with utter incredulity. I remember, a professor of Clark University wrote on that occasion and urged us to avoid such absurdities and thereby making the magazine ridiculous." (p. 22)

The articles in *McClure's* tackled a wide variety of scientific topics, from earthquakes to flying machines to astronomy in the interest of disseminating the latest ideas. Before the magazine even launched, McClure contacted many American scientists to contribute to a series call "The Edge of the Future." However, as Harold S. Wilson (1970) writes in *McClure's Magazine and the Muckrakers*, "the difficulty was that none of the men could write with any lucidity." (p. 116)

In McClure's mind, writes Wilson, "another scheme had been taking shape: to hire writers who could seize hold of new, complex, but interesting ideas, study them down to the bottom, and then put them in readable prose. Curiously, no editor before McClure had thought to do such a simple thing and, just as curiously, there were not many writers around who were equipped to meet the demand." (Wilson, 1970, p. 116)

So McClure hired professional writers Cleveland Moffett, Henry J.W. Dam, Ida Tarbell, and others who reported on scientific topics. They wrote clear, compelling articles, on themes ranging from the seemingly mundane (food, water, and wine inspection in Paris) to trendy (the effort to build a powered flying machine) to the unexpected (scientific kite flying). McClure's reporters also interviewed top scientists of the time, including Edison, Bell, and chemist and physicist James Dewar.

McClure did not entirely abandon scientists as writers. He solicited researchers to write about their work. In the first issue, Scottish evangelist and scientist Henry Drummond wrote on evolution, natural selection, and where humans got their ears. Sir Robert Ball, the royal astronomer of Ireland, later wrote about the possibility of life on other planets.

The science articles in *McClure's Magazine* were not confined to applied science and inventions like airships, dynamite, and automobiles. Henry Dam wrote about the search for absolute zero by Dewar. *McClure's Magazine* certainly did not make a distinction between basic and applied science as it ran stories from across the scientific and technical spectrum, with the only seeming criterion that they were interesting.

One good example of the intersection of basic and applied science was the discovery of X-rays by William Röntgen, which *McClure's* scooped in 1896. Röntgen had discovered a form of radiation that penetrates physical material and leaves a photographic record, an incomprehensible achievement to many reporters. McClure saw a story about X-rays in the newspaper "written with the heavy-handed ridicule which at the time journalists traditionally reserved for subjects they could not understand." (Wilson, p. 118) Attesting to the genius of McClure's approach to scientific topics as well as his understanding of the application of new ideas, the editor dispatched reporters to interview Röntgen and Professor Arthur Wright at Yale. They moved from Röntgen's revelation of his astounding discovery on January 4, 1896, to a twenty-page spread in the magazine by the April 1896 issue, an absurdly quick turnaround in that era.

The discovery of X-rays, of course, has implications to the present, as do Marconi's experiments in wireless transmissions, John Milne's work on seismography, Berthelot's studies in chemical synthesis of food, water quality inspections, and many of the other topics covered in *McClure's Magazine*. Those stories had a role in building public support for science. By bringing these ideas into the public imagination in a digestible way, the articles served a valuable role in diffusing knowledge, an essential part of any society's advancement. Harold Sharlin (1966) wrote in *The Convergent Century*,

> "The diffusion of science is just as serious a concern as the encouragement of science. … Diffusion means the spreading of scientific ideas, first among the creative experienced scientists and then a gradual extension. The next to be influenced by new scientific ideas are the part-time investigator and amateur. From this point science becomes popular and becomes public knowledge. Fully as important as scientific application is the absorption of scientific knowledge into the culture." (p. 182)

With a large subscription base and a well-connected, educated readership, *McClure's Magazine* spread new scientific thought through the minds of Americans in the late 1890s. It also represented the rise of professional science writing and writers who could perform the core task of science writing, which is what MIT professor of science writing Tom Levenson (2012) calls "the transformation of technically complicated material into a narrative available to broad audiences."

McClure believed in providing those scientific narratives to his readers, and succeeded through the strength of his magazine. Wilson (1970) does not exaggerate when he writes, "Judged from the standpoint of impact on its times, of the daring and vision of its editorial formula, of the sustained excellence of its editorial material, *McClure's* has never had a peer. ...McClure may, on the basis of his product, be confidently set down as the greatest magazine editor this country has yet produced." (p. 113)

Whether the articles surveyed automobile use, delved into dynamite factories, explained how volcanoes work, or described the uses of kites for scientific research, the stories were compellingly written and very informative. It was an exceptional publication in which a reader in the 1890s could learn about X-rays or evolution after reading a Robert Louis Stevenson story or an investigation of the Standard Oil Company's monopoly. The articles stand the test of time because, in the end, they are just good writing about science and the people behind the science and the inventions.

For modern readers, the articles and illustrations in this collection offer a window into the science of the late nineteenth century. With the rise of steampunk culture and a renewed interest in Victorian science, the articles curated in this volume give source materials directly from that era.

The prognostications of scientists of the time also give insight into the role of vision in scientific endeavors. Alexander Graham Bell's far-ranging imagination stoked future technical accomplishments, even if we can't quite read thoughts as he predicted. Berthelot's synthetic foods didn't create the utopia he postulates, but he nailed the necessity of dreaming for scientists when he said, "These are dreams, of course, but science may surely be permitted to dream sometimes. If it were not for dreams, where would be our impulse to progress?"

References

Daniels, G. H., editor. (1972). *Nineteenth-Century American Science: A Reappraisal*. Northwestern University Press.

Johnson, E. M., & Levenson, T. (2012). "The Uses of the Past: Why Science Writers Should Care About the History of Science—And Why Scientists Should Too." Scientific American blog, retrieved January 17, 2012. http://blogs.scientificamerican.com/primate-diaries/2012/01/17/uses-of-the-past/

Knight, D. (1976). *The Nature of Science: The History of Science in Western Culture since 1600*. Andre Deutsch, London.

McClure's Magazine. Vols. 1-13. June 1893-October 1899.

McClure, S.S. (1914). *My Autobiography*. Frederick A. Stokes Company, New York.

Sharlin, H. I. (1966) *The Convergent Century: The Unification of Science in the Nineteenth Century*. Abelard-Schuman, London.

Wilson, H. S. (1970). *McClure's Magazine and the Muckrakers*. Princeton University Press.

Inventions

The Maxim Air-Ship

An Interview with the Inventor.

By H. J. W. Dam

January 1894

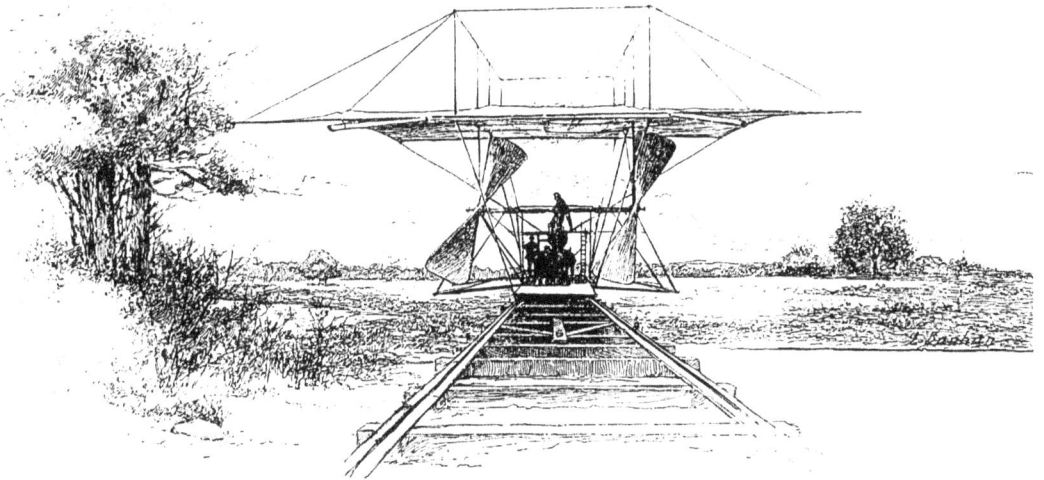

T HE fact that in a few years, probably ten at the outside estimate of the experts, the navigation of the air will be successfully and safely accomplished, may be news to many, though it is well understood by all who have followed the rapid development of applied science in this direction for some years past. Ever since the project of a dirigible balloon was abandoned by the foremost experimentalists in mechanical flight, in favor of an air-ship heavier than the air, and raised by mechanical power, a steady progress toward ultimately certain success has been clearly apparent. Very few people, however, are aware of the advanced results which have already been attained, and a visit to Baldwyn's Park, near Bexley, England, would be to them a revelation which can only be described as startling. To see a great air-ship, weigh-ing three and a half tons, flying across a park, on wheels, and to know that its engineer could lift it into the air, in a moment, by a turn of his wrist, makes one doubt the evidence of his own senses. It comes upon him with a shock, as if he had just awakened from a long Rip Van Winkle slumber, during which the magic of the world's advancement had left him hopelessly behind. The big white machine is a practical, moving fact, however. It can propel and lift it-self. And just as soon as those subordinate experiments, upon which depends the safety of aerial voyages, are completed, one of the greatest mechanical problems of the ages will have been finally and practically solved.

Among all the scientific men whose researches have contributed to this most important result, Mr. Hiram S. Maxim, the inventor of the air-ship in question, stands foremost. As the inventor of the Maxim gun, and many other ingenious machines of less importance, he had won a world-wide fame before the navigation of the air became the chief object of his study and investiga-tion. Beginning life fifty-three years ago, with a common-school education and a jack-knife, in Sangerville, Maine, he is now the proud possessor of a town house in London, and is lord of the manor at Baldwyn's Park, a stretch-ing domain of hundreds of acres, which he leased five years ago as well adapted to his preliminary experiments. Mr. Maxim is a man of medium height and solid build, his weight being two hundred and ten pounds. His hair, moustache, and beard are white, but his mental and physical energy are astonishing, and go far to explain the variety and extent of the results he has achieved. The work of inventing and constructing a flying-machine, near-ly every part of which, from boiler to connecting rods, is a variation from ex-isting appliances, enforced by the necessities of the occasion, is one which could only be undertaken by a man of much ingenuity, equipped with an ex-traordinary practical knowledge of mechanics. Even with these advantages, success would be impossible without unfailing energy and industry. All these qualities are, however, clearly visible in the manner and speech of the inven-tor. His voice and action show great physical strength, while his eyes, which are a deep brown, full and wide open, have continuously the semi-absorbed, preoccupied look of the student concentrated upon a problem. A courteous host, a jolly, even boisterous, story-teller, and a wonderful mechanician, Mr. Maxim is in his way as unusual as his machine. Withal he has a sturdy Americanism which personal interviews with half the reigning monarchs of Europe have not in the least affected, and he retains a pleasant conviction that of all the spots on the map of the world, not one is so important or so

The Workshop.

agreeable to contemplate as the good old "down-East" State of Maine. The American flag hangs in his hall, and he regards the United States as the safest, in fact the only, place in which to invest his money; a conclusion which is not without its importance, considering that his knowledge of European countries from the military, political, and financial standpoints has been attained through the channels of the gun business, and is, therefore, both comprehensive and exact.

For excellent and obvious reasons the visitors to Baldwyn's Park have been few. In his answer to my request for the privilege, Mr. Maxim wrote that he did not think it well to "say too much at present," but expressed his willingness to give "a few safe particulars." Upon arrival at Baldwyn's Park, the proceedings began with an appetizing luncheon in a handsome dining-room, though above the table hung an ornament which was perfectly calculated to entirely take away a visitor's appetite. This was a model of the air-ship on the scale of an inch to the foot. It was so strange an object that it made one oblivious of the fruit and indifferent to the coffee. By way of introduction and explanation the inventor said:

"The principle I have worked on, generally speaking, is that of the kite. That large cloth frame at the top of the model is the aeroplane, or main kite surface. The lesser aeroplane above the platform, or car; the side aeroplanes, or wings; and the flat-pointed rudders, fore and aft, are designed to furnish additional kite surface. It is necessary to make it, however, so that we can run it in a calm, against the air, thus making our own wind, as it were; and for this purpose I have a railway track, and instead of cords to hold the kite against the wind, I employ a pair of powerful screw propellers driven by a steam engine. In this manner I can drive the machine exactly as I please, can ascertain exactly how much the push of the screws is, and at the same time find out exactly how much the machine lifts at different speeds. The machine is, in fact, a big kite. Should I fly it in the air with a cord during a strong gale and then run my engines, I should be able to find out how fast they would have to run in order to take all the pull off the cord. As soon as the cord became slack, the machine would be flying with its own engine power."

To more clearly illustrate his meaning, Mr. Maxim led the way to the workshop in the grounds—a large and substantial bird-cage, sixty feet wide and fifty high, in which the mechanical bird had been constructed, and stood perched for one of its daily flights. A railway track, nine feet wide, ran outward from the closed doors, and stretched indefinitely, in a straight line, across the green level of the park to the line of a belt of woods two thousand feet distant. The front of the shop consisted of four large doors, "the largest in the world," their owner remarked; and when these were rolled back by a dozen workmen the air-ship came into view. It was so novel, so unexpected, and so apparently complex at first sight, that it held the eye for a long, silent period; the beholder's sensation being one of wonder, if not awe, coupled with an indescribable mechanical confusion of ideas.

It took many minutes to grasp it; to form an intelligent idea of it. Then, as the sense of relation between the different parts developed, it became a framework of black steel rods of varying sizes, with a square frame of white cloth, fifty feet by fifty at the top, and an inclined wooden platform, eight feet wide by forty long, resting on wheels upon the track below. On the platform, near the front end, was a small boiler-house in the shape, roughly speaking, of a truncated pyramid, and ten feet behind it was a frame eleven feet high, on which were two sets of compound cylinders, and two big wooden screws above the two sides of the platform, and eighteen feet apart. Outside of these fundamental accessories were a water-tank, a naphtha-tank, and an indefinite number of rods and very small wire ropes, to give strength and compactness to the whole. The many minor elements of the machinery did not at first catch the eye, but all appeared in interesting action when details were entered upon later on. It should be noted that the machine, as it stood and as it appears in the accompanying pictures, was without the side planes, and the big rudders of cloth on steel frames, which are mounted, fore and aft, on the main aeroplane. These are not used in the experimental trials, their utility having been established, as far as is possible without a practical test in the air.

Pushed by the workmen, the machine rolled slowly out of the house, and shortly stood upon the track in the park. It had completely filled the

The air-ship in the workshop.

workshop from roof to floor; but here, with only the sky above it, seemed smaller and lighter. The steam was hissing in the boiler; the big screws had made one or two preliminary revolutions, and a flight along the track was imminent. "Jump on board," shouted its owner, who stood at the boiler, conning half a dozen different gauges; and, climbing over an outlying rod like the outrigger of a canoe, I mounted the platform, which was of the lightest matched boards, so thin that they seemed insufficient to bear a man's weight. Prior to the start, a rope running to a dynamometer and post was attached behind to measure the forward impulse, or "push," of the screw. Mr. Maxim turned on the steam, and the screw on the port side began to revolve. It is seventeen feet eleven inches in length, five feet wide at the ends, and twenty-two inches at the waist. It is made of the lightest American yellow pine, and painted a pale blue, the paint having been sandpapered to perfect smoothness, reducing the skin, friction to a point at which it became negligible. It revolved faster and faster as the steam power was increased, until it was whirling on its seemingly frail framework at a dizzying speed. Then steam was shut off; it came quickly to a standstill, and its fellow on the other side was tried. All working smoothly, both screws began to turn faster and faster and faster, until the eye began to lose the blades and retain only the sense of two whirling discs. The action of the screws at high speed caused remarkably little shaking of the whole machine. This is one of the surprises of the invention, the tremendous force exerted as compared with the lightness, steadiness, and compactness of the whole.

Behind the screws, forty feet away, two men were squatting over the dynamometer, and indicating the degree of "push" on a large index board for the engineer to read. The index marked four hundred, live hundred, six hundred, seven hundred, and, finally, twelve hundred pounds of "push." The pressure was then diminished below five hundred, and the commander yelled: "Let go." A rope was pulled, the machine shot forward like a railway train, and, with the big wheels whirling, the steam hissing, and the waste pipes puffing and gurgling, flew over the eighteen hundred feet of track in much less time than it takes to tell it. It was stopped by a couple of ropes stretched across the track, working on capstans fitted with revolving fans. The stoppage was gentle, and the passenger breathed freely again, looking now upon the machine with more friendly and less fearful eye, as if it were a dangerous bulldog with which amicable relations had been established and fear of injury was over. The machine was then pushed back over the track, it not being built, any more than a bird, to fly backward. In a quarter of an hour it is again at its

The air-ship on the track.

starting place, and ready for another flight. Having seen it in action and had evidence of its power, the details were more than ever interesting, and were furnished by the inventor in succinct and practical terms.

The first question was its supporting power in the air. He said:

"The area of the main aeroplane is two thousand eight hundred and nine-ty-four square feet; of the small one, one hundred and twenty-six; and of the bottom of the car, one hundred and forty. With the rudders and wings added, the total area is about six thousand square feet. The wings are ten in number, and superposed, five on each side, and are each five feet wide and from twenty-five to thirty-five feet in length, according to their positions. he forward rudder, projecting in front from the main aeroplane, is eighteen feet wide and thirty feet long, and the aft one, eighteen by twenty-three. Rudders and wings, like all the other aeroplanes, are made of a specially woven cotton cloth, so fine that you cannot blow through it, and mounted on a framework of hollow steel tubes. All these aeroplanes are inclined at a small angle to the air, the angle which gives the most support combined with the least resistance to its forward motion."

"What speed is necessary to support the machine in air?"

"A minimum, under present conditions, of twenty-five miles an hour. At that speed, with wings and rudders adjusted, it will leave the track. It lifted in one of the earlier trials, and caused us some trouble, as we were not ready."

"What will happen in the air if anything goes wrong, and the engine stops?"

"The machine will settle to the earth, and land with the same velocity as if it had fallen a distance of three feet."

"Only three feet?"

"Yes. When the propulsion ceases, the machine will fall three feet. At this point the resistance to the atmosphere afforded by the aeroplanes will become nearly equal to the force of gravity, and it will settle without any increase of velocity."

"How about its steadiness in the air? You know a kite sometimes indulges in extraordinary rolling, to say nothing of darts and dives."

The explanation of this point was given ocularly, and much more clearly than words would have made it. Mr. Maxim tore a sheet of paper from his note-book, held it up, and let it fall to the ground. It darted, dived, and fell in irregular lines, shooting out behind him. He then took the same sheet of paper, tore a square out of each corner, and bent back the four sides from the corners of the squares at an angle of forty-five degrees. He then held this up and let it fall. It sank to the earth gently, without a tremor, its surface remaining perfectly even throughout. "That," said he, "is the principle of the wings. They are so adjusted that as any side is depressed it presents a greater lifting surface to the air below. There's no trouble keeping her on an even keel," he added, with a smile.

"But can't it tip over in a wind?"

"No. It is quite possible to make a plane remain right side up in the air, even if the centre of gravity is considerably above the centre of lifting effort. Stability in the air depends very largely upon the shape of the aeroplane, but nevertheless with this machine the centre of gravity is very much below the centre of lift; and this, together with the form of the aeroplane, makes it quite impossible that the machine should tip over in the air. The centre of gravity in this machine is here," and he held up his hand at an imaginary point about five feet back of the boiler, and seven feet above the centre of the platform. It may be here mentioned that the main aeroplane is twenty-five feet above the platform. The total height of the machine to the tops of the rods above the aeroplane is thirty-five feet, and its greatest length seventy feet.

"Are the cotton aeroplanes strong enough to bear the weight in falling, without fracture?"

"They are twenty-five times stronger than is necessary. The greatest weight which can bear on them is a little over a pound to the square foot, and they are tested for twenty-five pounds. The pressure on the cloth is practically the same at all speeds, whether the machine is falling to the earth or sailing through the air; the cloth in any case has to sustain the weight of the machine."

"How is it steered?"

"For steering to the right or left I expect to use the screws. If I have any difficulty I can easily use rudders. For steering upward or downward the fore and aft rudders will be used. The aft one is pivoted on the extension of the two centre poles, and the forward one hung on their ends. Both will be worked from the centre of the platform, and will at first require a man to each, though I shall greatly simplify the working of them later on."

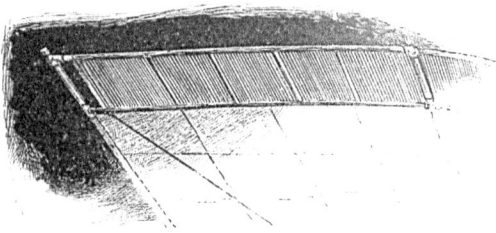

Condensing tubes on the edge of the aeroplane.

"What is your estimate of the speed?"

"I don't expect, with this machine, to get over thirty-five miles per hour. The next one, which will be smaller, and will be worked with a hundred horse-power, will give me, I expect, from fifty to sixty miles per hour. The highest speed I look for, as the art is perfected, is ninety miles per hour. I believe that any speed which is attained by a railway train can be reached by a machine moving through the air."

"How about the duration of the flight?"

"That is merely a matter of water and naphtha. The margin of weight-carrying is so large that, once the machine is successful, any amount of time and distance within reason can be looked for."

As far as support and action in the air were concerned, there seemed nothing more to be said, and yet it was difficult to realize that the facts as stated were simply and undeniably true; to realize that the navigation of the air is the traversing of an entirely new medium, whose conditions are so foreign to those of water, for instance, that they are difficult to quickly conceive.

The next question was that of weight, and here came some object lessons in the weight of metal that were astonishing. "Lift that tube," said Mr. Maxim. The tube was of copper, four feet long, and elliptical in shape, its greatest diameter being one and a half inches. It looked heavy. Lifted up, its lightness was surprising. It weighed no more than thin paper, and actually seemed, for the moment, like paper colored in imitation of copper. "That is one of the condensing tubes," said Mr. Maxim. "There are five hundred of them up there," and he pointed to a section of what had appeared to be thin laths running across the entire front of the main aeroplane. "Of course," said he, "we can't waste any water up in the air, because we have no means of replenishing. The used steam runs up by those large pipes, and the water runs back through those small ones to the tank in the centre of the platform. The framework is constructed," he continued, "not of rods, but tubes, and tubes of the least possible weight. They are all of steel, a steel with considerable carbon in it and not tempered, and they vary from one inch to three inches in diameter. I tried aluminium, but found that steel was stronger, weight for weight. In addition to this, steel tubes can be united with great facility, and the coefficient of the joint is fully ninety-five. There is no convenient way of uniting aluminium tubes, however, and if they were united the coefficient of the joint would be very low. The heaviest tubes in the machine are the shafts of the screws, which are five inches in diameter, five feet long, and an eighth of an inch thick. The next size, used in the car, are three inches in diameter, and one twelfth of an inch thick. I have a few more, one fourteenth of an inch thick, of the same size. I need not say that at every point I have used the lightest tube possible for the strain which comes upon it, perfect safety being at all times considered, as I purpose to take my first machine up into the air myself, and I don't intend to run any risks. The bulk of the machine is constructed of hard steel tubes one twenty-fifth of an inch in thickness. The total weight of the machine, with its full complement of water, naphtha, and three men, is something over seven thousand one hundred pounds. Without the wings it is six thousand eight hundred and eighty. The boiler complete weighs one thousand pounds. This small weight, considering it gives me a force of three hundred horse-power, is perhaps the most valuable portion of the work, since it has always been known that we could fly if we could get a motive power of adequate strength with sufficient lightness. I use a compound engine, the high pressure cylinders being five inches in diameter, with a twelve-inch stroke; and the low pressure, eight inches in diameter, with a twelve-inch stroke. The piston speed is eight hundred feet per minute. Nearly everything connected with the machinery had to be newly designed, with a special view to lightness, none of the known appliances being of use in this case. It was necessary, in the first place, to develop a system of making a very large quantity of carburetted air from naphtha, with very little weight."

Pointing out a large hole where the air was drawn in, he said, that, as the velocity with which the combined air and gases entered was at the rate of two miles a minute, he found it very difficult to deal with these gases at this high velocity, and had spent a great deal of time in devising a system by which the gas was equally spread out over the whole furnace, and not influenced by the inductive action of the incoming gas at this very high velocity. "I had," he resumed, "to devise a system for regulating the product of the gas; for pumping the liquid into the gas generator; a new kind of boiler and feed-water heaters; a system for burning a very large quantity of carburetted air in a small space, without smoking or blowing out; a system for regulating the steam, and pumps for filling the boiler and regulating the supply. None of the existing types of engines seemed well fitted to the purpose. I had to design one expressly with a view to great lightness; and, notwithstanding there were some hundreds of types of connecting rods already in existence, I found it necessary to design an absolutely new form of connecting rods. I had to invent a new dynamometer to meet the necessities, and new dynagraphs for measuring the lift of the machine at different speeds, as well as another to measure its rate of speed through the air." He paused, looking over at the machine which represented so many hours of concentrated brain work in a puzzled, absorbed way. "And there's more to do yet," he added impressively. "I don't call this an air-ship or a flying-machine or anything else. To me it is merely a machine for making experiments in aerial navigation. In my next one, I shall make a number of changes which it is not worth while to make in this. It is slow work, but there is no doubt of the result. Propulsion and lifting are solved problems, and it is merely a matter of time."

"How much time?"

Mr. Maxim illustrating the principles of the wings of the air-ship.

"Well, if I had nothing else to occupy me, unlimited money, and plenty of space for experimenting, I should expect to be up in the air within eighteen months. I am very busy, however, have a very limited space here, and am proceeding as economically as possible. In my opinion, however, under the most unfavorable conditions, aerial navigation will be an accomplished fact inside of ten years."

This was a digression. We now came back to the most remarkable boiler that ever was seen. It was enclosed in a house eight feet long, five feet wide at the base, and about six feet high. The sides of the house were of thick cloth, woven from pure asbestos, and the frame and top of the thinnest iron. Within, viewed through a peephole, the entire floor was a mass of small flames from seven thousand six hundred gas burners. The boiler has about six hundred tubes which are eight feet long, and about one hundred which are four feet ten inches long. These tubes are about half an inch external diameter, and half a millimeter, or one-fiftieth of an inch, in thickness. They are curved and joined into a steam drum, ten inches in diameter and eight feet long, where the water and steam are separated, the water again passing through the boiler, and the steam passing to the engine. There are also some three or four hundred much smaller tubes, which are used for heating the water by the products of combustion before it enters the main boiler at all. In order to prevent the tubes from being injured by the great heat of the fire, a forced circulation of the water is employed. It is therefore possible to use a very small and thin tube and a very hot fire without any danger. A single spare boiler tube in the shop served to exhibit the peculiar lightness of the boiler, which is perhaps the most ingenious, as well as the most important, part of the machine. The tube, like the condensing tube before mentioned, was as light as so much paper. It was made of pure copper, any impurities, in view of the thinness of the tubes, causing them to become "hot short" and break. "With only a moderate fire," said Mr. Maxim, "I have been able to get a horse-power out of four of these tubes; with a *hotter* fire I have gotten a horse-power out of three of them. Their bursting pressure under steam is sixteen hundred and fifty pounds to the square inch. The boiler itself has been fired to give a steam pressure of four hundred and ten pounds to the square inch, but I have never run the engine above three hundred pounds, thereby developing three hundred brake horse-power, which is all that I need for this weight, and which leaves a very wide margin of safety. To run the boiler the machine carries six hundred pounds of water, and two hundred pounds of seventy degree Beaumé naphtha. The consumption of naphtha is about one

Details of the Maxim air-ship.

pound per horse-power per hour."

Last of all, in the way of general description, came the questions of propulsion and lifting power. To give all the details, under this heading, into which the inventor entered, would alone make an article quite as long as this, if not a small volume. Concerning specific results, however, he said:

"The lifting of an aeroplane by a screw or screws has been the subject of many series of experiments by myself and others. The number of pounds lifted by one pound of 'push' in the screw varies greatly with conditions. In my early experiments with a merry-go-round, or whirling table, I succeeded in lifting fourteen times the 'push' of the screw, or fourteen pounds of weight for every pound of 'push' forward. In this large machine, however, with a large number of wires and a good deal of framework, where the aeroplane is so large, where it is difficult to make it remain uniform or rigid when there is a pressure on it, and where I have an engine, boiler, platform, men, tanks, wires, and tubes to force through the air, I have not been able to lift more than six pounds for each pound of 'push.' This, however, is much more than is absolutely necessary. The engine is able to give, and has often given, a 'push' of nineteen hundred and sixty pounds, which would mean a lifting power of nearly twelve thousand pounds. With a push of one thousand pounds from the screws, using one hundred and twenty horse-power, the lift, as shown by the dynagraphs, was over six thousand pounds. This left only a weight of one thousand pounds on the track, and this was not sufficient to keep us there. The speed along the track, with this 'push,' was twenty-seven miles per hour."

"When do you expect to take your first flight?"

"I have not set any time, and shall not. Haste in an enterprise of this kind is the worst possible policy. At every trial of a machine which is mechanically new in so many particulars, weak points develop and require attention, while new improvements constantly suggest themselves. To-day it is a leaking valve, to-morrow something else. Rising into the air with a new machine, when all the experiments in the way of manœuvering, which can only take place in the air, are yet untried, would be unwise until everything which can be completely tested on the track has been so tested. The possibilities of accident must be as nearly as possible exhausted beforehand. More than this, I have not at Baldwyn's Park the necessary room and privileges. It may be that I shall not attempt to rise until I have more room, and I am now looking for a suitable location—something difficult to find in England. In fact," he added, with one of his ready New England comparisons, "I'm like a boy with a pair of skates which he has never tried, and only a little piece of ice to try them on."

The foregoing was the substance of the "few safe particulars" which Mr. Maxim was willing to give. The improvements upon his first machine, which will appear in the second, and the eventualities and possibilities of aerial navigation, were subjects upon which he was not inclined to talk very much. He confessed, however, that an air voyage of three or four thousand miles seemed to him eventually probable. "I don't want to speak of things before I am ready to do them. I don't imagine that flying-machines will be used very soon to carry bricks from Haverstraw to New York, or coals from Newcastle. The first machines are certain to be used for military purposes, whatever their cost or whatever the expense of running them, and the nation which first employs them will have every other at its mercy. I shall be quite content with my results when I can go a distance of twenty miles and back. That will suffice for all present purposes."

> *Note.*—The illustrations for this article are from photographs
> taken under the supervision of the author and Mr. Maxim, by
> Pradelle & Young of Regent Street, London.

Marconi's Wireless Telegraph

Messages Sent at Will Through Space. —Telegraphing Without Wires Across the English Channel.

By Cleveland Moffett

June 1899

NOTE.—All the illustrations in this article, except the picture of the Royal Yacht "Osborne," were made expressly for *McClure's Magazine.*

MR. **MARCONI** began his endeavors at telegraphing without wires in 1895, when in the fields of his father's estate at Bologna, Italy, he set up tin boxes, called "capacities," on poles of varying heights, and connected them by insulated wires with the instruments he had then devised—a crude transmitter and receiver. Here was a young man of twenty hot on the track of a great discovery, for presently he is writing to Mr. W. H. Preece, chief electrician of the British postal system, telling him about these tin boxes and how he has found out that "when these were placed on top of a pole two meters high, signals could be obtained at thirty meters from the transmitter;" and that "with the same boxes on poles four meters high signals were obtained at 100 meters, and with the same boxes at a height of eight meters, other conditions being equal, nearly up to a mile and a half. Morse signals were easily obtained at 400 meters." And so on, the gist of it being (and this is the chief point in Marconi's present system) that the higher the pole (connected by wire with the transmitter), the greater was found to be the distance of transmission.

In 1896, Marconi came to London and conducted further experiments in Mr. Preece's laboratory, these earning him followers and supporters. Then came the signals on Salisbury Plain through house and hill, plain proof for doubters that neither brick walls nor rocks nor earth could stop these subtle waves. What kind of waves they were Marconi did not pretend to say; it was enough for him that they did their business well. And since they acted best with wire supported from a height, a plan was conceived of using balloons to hold the wires, and March, 1897, saw strange doings in various parts of England: ten-foot balloons covered with tin-foil sent up for "capacities" and promptly blown into slivers by the gale; then six-foot calico kites with tin-foil over them and flying tails; finally tailless kites, under the management of experts. In these trials, despite unfavorable conditions, signals were transmitted through space between points over eight miles apart.

South Foreland, the English station from which messages were sent without wires to Boulogne, France, thirty-two miles away. The mast supporting the vertical wire is seen on the edge of the cliff.

In November, 1897, Marconi and Mr. Kemp rigged up a stout mast at the Needles on the Isle of Wight, 120 feet high, and supported a wire from the top by an insulated fastening. Then, having connected the lower end of this wire with a transmitter, they put out to sea in a tugboat, taking with them a receiving-instrument connected to a wire that hung from a sixty-foot mast. Their object was to see at what distance from the Needles they could get signals. For months, through storm and gale, they kept at this work, leaving the Needles farther and farther behind them as details in the instruments

24

were improved, until by the New Year they were able to get signals clear across to the mainland. Forthwith a permanent station was set up there—first at Bournemouth, fourteen miles from the Needles, but subsequently moved to Poole, eighteen miles.

An interesting fact may be noted, that on one occasion, soon after this installation, Mr. Kemp was able to get Bournemouth messages at Swanage, several miles down the coast, by simply lowering a wire from a high cliff and connecting on a receiver at the lower end. Here was communication established with only a rough precipice to serve and no mast at all.

Let us come now to the Kingstown regatta, which took place in July, 1898, and lasted several days. The "Daily Express" of Dublin set a new fashion in newspaper methods by arranging to have these races observed from a steamer, the "Flying Huntress," used as a movable sending-station for Marconi messages which should describe the different events as they happened. A height of from seventy-five to eighty feet of wire was supported from the mast, and this was found sufficient to transmit easily to Kingstown, even when the steamer was twenty-five miles from shore. The receiving-mast erected at Kingstown was 110 feet high, and the despatches as they arrived here through the receiving-instrument were telephoned at once to Dublin, so that the "Express" was able to print full accounts of the races almost before they were over, and while the yachts were out far beyond the range of any telescope. During the regatta more than 700 of these wireless messages were transmitted.

Not less interesting were the memorable tests that came a few days later, when Marconi was called upon to set up wireless communication between Osborne House, on the Isle of Wight, and the royal yacht, with the Prince of Wales aboard, as she lay off in Cowes Bay. The Queen wished to be able thus to get frequent bulletins in regard to the Prince's injured knee, and not less than 150 messages of a strictly private nature were transmitted, in the course of sixteen days, with entire success. By permission of the Prince of Wales, some of these messages have been made public; among others the following:

August 4th.

From Dr. Tripp to Sir James Reid.

H. R. H. the Prince of Wales has passed another excellent night and is in very good spirits and health. The knee is most satisfactory.

August 5th.

From Dr. Tripp to Sir James Reid.

H. R. H. the Prince of Wales has passed another excellent night, and the knee is in good condition.

The transmission here was accomplished in the usual way—with a 100-foot pole at Ladywood Cottage, in the grounds of Osborne House, supporting the vertical conductor, and a wire from the yacht's mast lifted eighty-three feet above deck. This wire led down into the saloon, where the instruments were operated and observed with great interest by the various royalties aboard, notably the Duke of York, the Princess Louise, and the Prince of Wales himself. What seemed to amaze them above all was that the sending could go on just the same while the yacht was plowing along through the waves. The following was sent on August 10th by the Prince of Wales while the yacht was steaming at a good rate off Benbridge, seven or eight miles from Osborne:

William Marconi.
From a photograph taken especially for *McClure's Magazine* at South Foreland Lighthouse, March 29, 1899

To the Duke of Connaught.

Will be pleased to see you on board this afternoon when the "Osborne" returns.

On one occasion the yacht cruised so far west as to bring its receiver within the influence of the transmitter at the Needles, and here it was found possible to communicate successively with that station and with Osborne, and this despite the fact that both stations were cut off from the yacht by considerable hills, one of these, Headon Hill, rising 314 feet higher than the vertical wire on the "Osborne."

It was at the extreme west of the Isle of Wight that I got my first practical notion of how this amazing business works. Looking down from the high ground, a furlong beyond the last railway station, I saw at my feet the horseshoe cavern of Alum Bay, a steep semicircle, bitten out of the chalk cliffs, one might fancy, by some fierce sea-monster, whose teeth had snapped in the effort and been strewn there in the jagged line of Needles. These gleamed up white now out of the waves, and pointed straight across the Channel to the mainland. On the right were low-lying reddish forts, waiting for some enemy to dare their guns. On the left, rising bare and solitary from the highest hill of all, stood the granite cross of Alfred Tennyson, alone, like the man, yet a comfort to weary mariners.

Here, overhanging the bay, is the Needles Hotel, and beside it lifts one of Mr. Marconi coni'stall masts, with braces and halyards to hold it against storm and gale. From the peak hangs down a line of wire that runs through a window into the little sending-room, where we may now see enacted this mystery of talking through the ether. There are two matter-of-fact young men here who have the air of doing something that is altogether simple. One of them stands at a table with some instruments on it, and works a black-handled key up and down. He is saying something to the Poole station, over yonder in England, eighteen miles away.

"Brripp—brripp—brripp—brrrrrr.
Brripp—brripp—brripp—brrrrrr—
Brripp—brrrrrr—brripp. Brripp—brripp!"

So talks the sender with noise and deliberation. It is the Morse code working—ordinary dots and dashes which can be made into letters and words, as everybody knows. With each movement of the key bluish sparks jump an inch between the two brass knobs of the induction coil, the same kind of coil and the same kind of sparks that are familiar in experiments with the Roentgen rays. For one dot, a single spark jumps; for one dash, there comes a stream of sparks. One knob of the induction coil is connected with the earth, the other with the wire hanging from the masthead. Each spark indicates a certain oscillating impulse from the electrical battery that actuates the coil; each one of these impulses shoots through the aërial wire, and from the wire through space by oscillations of the ether, travelling at the speed of light, or seven times around the earth in a second. That is all there is in the sending of these Marconi messages.

"I am giving them your message," said the young man presently, "that you will spend the night at Bournemouth and see them in the morning, Anything more?"

"Ask them what sort of weather they are having," said I, thinking of nothing better.

"I've asked them," he said, and then struck a vigorous series of V's, three dots and a dash, to show that he had finished.

"Now I switch on to the receiver," he explained, and connected the aërial wire with an instrument in a metal box about the size of a valise. "You see the aërial wire serves both to send the ether waves out and to collect them as they come through space. Whenever a station is not sending, it is connected to receive."

"Then you can't send and receive at the same time?"

"We don't want to. We listen first, and then talk. There, they're calling us. Hear?"

Inside the metal box a faint clicking sounded, like a whisper after a hearty tone. And the wheels of a Morse printing-apparatus straightway began to turn, registering dots and dashes on a moving tape.

"They send their compliments, and say they will be glad to see you. Ah, here comes the weather: 'Looks like snow. Sun is blazing on us at present.'"

It is worthy of note that, five minutes later, it began to snow on our side of the Channel.

"I must tell you," went on my informant, "why the receiver is put in this metal box. It is to protect it against the influence of the sender, which, you observe, rests beside it on the table. You can easily believe that a receiver sensitive enough to record impulses from a point eighteen miles away might be disorganized if these impulses came from a distance of two or three feet. But the box keeps them out."

Mast and station at South Foreland, near Dover, England, used by Mr. Marconi in telegraphing without wires across the Channel to Boulogne, France. *From a photograph.*

"And yet it is a metal box?"

"Ah, but these waves are not conducted as ordinary electric waves are. These are Hertzian waves, and good conductors for every-day electricity may be bad conductors for them. So it is in this case. You heard the receiver work just now for the message from Poole, yet it makes no sound while our own sender is going. But look here, I will show you something."

He took up a little buzzer with a tiny battery, such as is used to ring electric bells. "Now listen. You see, there is no connection between this and the receiver." He joined two wires so that the buzzer began to buzz, and instantly the receiver responded, dot for dot, dash for dash.

"There," he said, "you have the whole principle of the thing right before you. The feeble impulses of this buzzer are transmitted to the receiver in the same way that the stronger impulses are transmitted from the induction coil at Poole. Both travel through the ether."

"Why doesn't the metal box stop these feeble impulses as it stops the strong ones of your own sender?"

"It does. The effect of the buzzer is through the aërial wire, not through the box. The wire is connected with the receiver now, but when we are sending, it connects only with the induction coil, and the receiver, being cut off, is not affected."

"Then no message can be received when you are sending?"

"Not at the very instant. But, as I said, we always switch back to the receiver as soon as we have sent a message; so another station can always get us in a few minutes. There they are again."

Once more the receiver set up its modest clicking.

"They're asking about a new coherer we're putting in," he said, and proceeded to send the answer back. I looked out across the water, which was duller now under a gray sky. There

Mast and station at Boulogne, France, used by Mr. Marconi in telegraphing without wires across the Channel to South Foreland, England. *Drawn from a photograph.*

was something uncanny in the thought that my young friend here, who seemed as far as possible from a magician or supernatural being, was flinging his words across this waste of sea, over the beating schooners, over the feeding cormorants, to the dim coast of England yonder.

"I suppose what you send is radiated in all directions?"

"Of course."

"Then any one within an eighteen-mile range might receive it?"

"If they had the proper kind of a receiver." And he smiled complacently, which drew further questions from me, and presently we were discussing the relay and the tapper and the twin silver plugs in the neat vacuum tube, all essential parts of Marconi's instrument for catching these swift pulsations in the ether. The tube is made of glass, about the thickness of a thermometer tube and about two inches long. It seems absurd that so tiny and simple an affair can come as a boon to ships and armies and a benefit to all mankind; yet the chief virtue of Marconi's invention lies here in this fragile coherer. But for this, induction coils would snap their messages in vain, for none could read them. The silver plugs in this coherer are so close together that the blade of a knife could scarcely pass between them; yet in that narrow slit nestle several hundred minute fragments of nickel and silver, the finest dust, siftings through silk, and these enjoy the strange property (as Marconi discovered) of being alternately very good conductors and very bad conductors for the Hertzian waves—very good conductors when welded together by the passing current into a continuous metal path, very bad conductors when they fall apart under a blow from the tapper. One end of the coherer is connected with the aërial wire, the other with the earth and also with a home battery that works the tapper and the Morse printing-instrument.

And the practical operation is this: When the impulse of a single spark comes through the ether down the wire into the coherer, the particles of metal cohere (hence the name), the Morse instrument prints a dot, and the tapper strikes its little hammer against the glass tube. That blow decoheres the particles of metal, and stops the current of the home battery. And each successive impulse through the ether produces the same phenomena of coherence and decoherence, and the same printing of dot or dash. The impulses through the ether would never be strong enough of themselves to work the printing-instrument and the tapper, but they are strong enough to open and close a valve (the metal dust), which lets in or shuts out the stronger current of the home battery—all of which is simple enough after some one has taught the world how to do it.

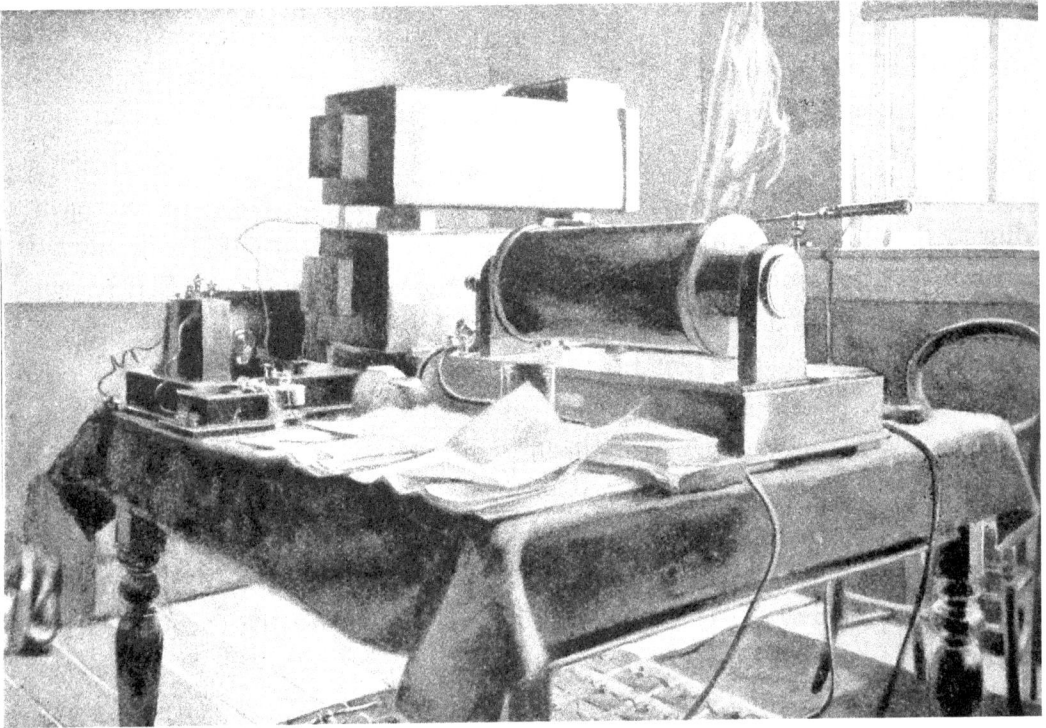

The apparatus employed at South Foreland lighthouse for communicating with the *Goodwin Sands* lightship and with Boulogne. *Drawn from a photograph.*

Twenty-four hours later, after a breezy ride across the Channel on the self -reliant side-wheeler "Lymington," then an hour's railway journey and a carriage jaunt of like duration over gorse-spread sand dunes, I found myself at the Poole Signal Station, really six miles beyond Poole, on a barren promontory. Here the installation is identical with that at the Needles, only on a larger scale, and here two operators are kept busy at experiments, under the direction of Mr. Marconi himself and Dr. Erskine-Murray, one of the company's chief electricians. With the latter I spent two hours in profitable converse. "I suppose," said I, "this is a fine day for your work?" The sun was shining and the air mild.

"Not particularly," said he. "The fact is, our messages seem to carry best in fog and bad weather. This past winter we have sent through all kinds of gales and storms without a single breakdown."

"Don't thunder-storms interfere with you, or electric disturbances?"

"Not in the least."

"How about the earth's curvature? I suppose that doesn't amount to much just to the Needles?"

"Doesn't it though? Look across, and judge for yourself. It amounts to 100 feet at least. You can only see the head of the Needles lighthouse from here, and that must be 150 feet above the sea. And the big steamers pass there hulls and funnels down."

"Then the earth's curvature makes no difference with your waves?"

"It has made none up to twenty-five miles, which we have covered from a ship to shore; and in that distance the earth's dip amounts to about 500 feet. If the curvature counted against us then, the messages would have passed some hundreds of feet over the receiving-station; but nothing of the sort happened. So we feel reasonably confident that these Hertzian waves follow around smoothly as the earth curves."

"And you can send messages through hills, can you not?"

"Easily. We have done so repeatedly."

"And you can send in all kinds of weather?"

"We can."

"Then," said I after some thought, "if neither land nor sea nor atmospheric conditions can stop you, I don't see why you can't send messages to any distance."

Transmitting instrument at Boulogne station. *Drawn from a photograph.*

"So we can," said the electrician, "so we can, given a sufficient height of wire. It has become simply a question now how high a mast you are willing to erect. If you double the height of your mast, you can send a message four times as far. If you treble the height of your mast, you can send a message nine times as far. In other words, the law established by our experiments seems to be that the range of distance increases as the square of the mast's height. To start with, you may assume that a wire suspended from an eighty-foot mast will send a message twenty miles. We are doing about that here."

"Then," said I, multiplying, "a mast 160 feet high would send a message eighty miles?"

"Exactly."

"And a mast 320 feet high would send a message 320 miles; a mast 640 feet high would send a message 1,280 miles; and a mast 1,280 feet high would send a message 5,120 miles?"

"That's right. So you see if there were another Eiffel Tower in New York, it would be possible to send messages to Paris through the ether and get answers without ocean cables."

"Do you really think that would be possible?"

"I see no reason to doubt it. What are a few thousand miles to this wonderful ether, which brings us our light every day from millions of miles?"

"Do you use stronger induction coils," I asked, "as you increase the distance of transmission?"

"We have not up to the present, but we may do so when we get into the hundreds of miles. A coil with a ten-inch spark, however, is quite sufficient for any distances under immediate consideration."

After this we talked of improvements in the system made by Mr. Marconi as the result of experiments kept up continuously since these stations were established, nearly two years ago. It was found that a horizontal wire, placed at whatever height, was of practically no value in sending messages; all that counts here is the vertical component. Also that it is better to have the wire conductor suspended out from the mast by a sprit. It was found, furthermore, that by modifying the coherer and perfecting various details of installation the total efficiency was much increased, so that the vertical conductor could be lowered gradually without disturbing communication. Now they are sending to the Needles with a sixty-foot conductor, whereas at the start a wire with 120 feet vertical height was necessary.

So much for my visits to these pioneer etherial stations (if I may so style them), which gave me a general familiarity with the method of wireless

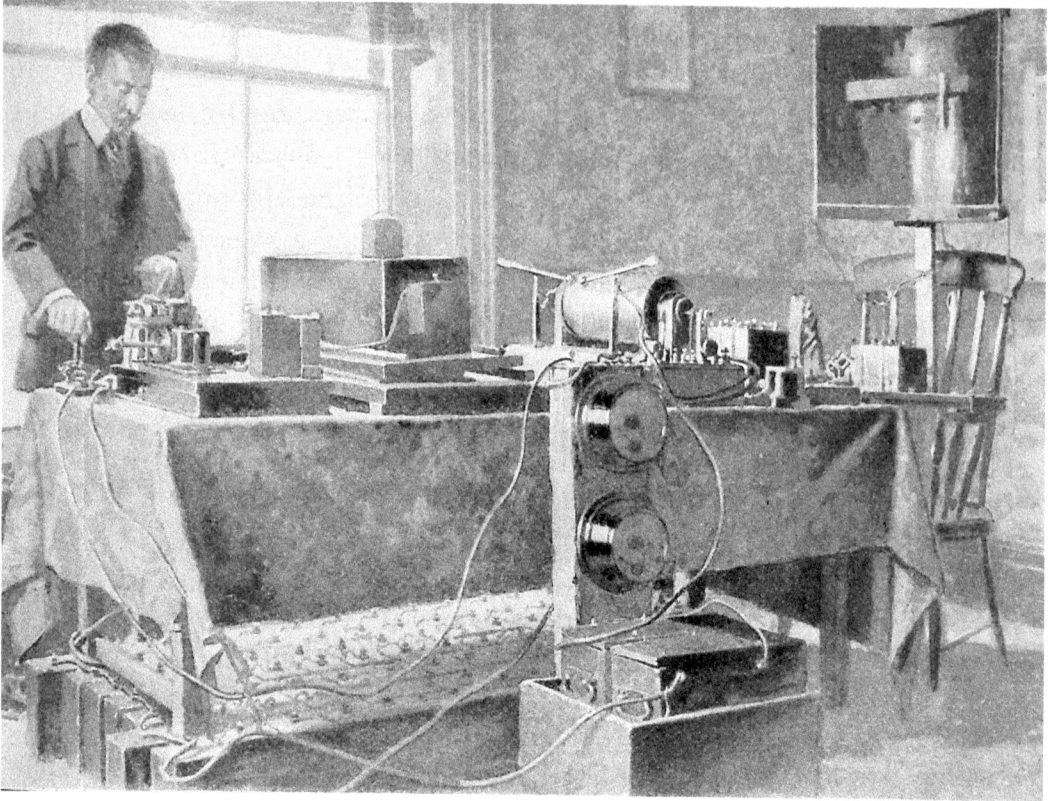

The wireless telegraph station at Poole, showing sending and receiving instruments. In the right-hand corner is the copper reflector used in directing the waves. *Drawn from a photograph.*

telegraphy and enabled me to question Mr. Marconi with greater pertinence during several talks which it was my privilege to have with him. What interested me chiefly was the practical and immediate application of this new system to the world's affairs. And one thing that came to mind naturally was the question of privacy or secrecy in the transmission of these aërial messages. In time of war, for instance, would communications between battleships or armies be at the mercy of any one, including enemies, who might have a Marconi receiver?

On this point Mr. Marconi had several things to say. In the first place, it was evident that generals and admirals, as well as private individuals, could always protect themselves by sending their despatches in cipher. Then, during active military operations, despatches could often be kept within a friendly radius by lowering the wire on the mast until its transmitting power came within that radius.

Marconi realizes, of course, the desirability of being able in certain cases to transmit messages in one and only one direction. To this end he has conducted a special series of experiments with a sending-apparatus different from that already described. He uses no wire here, but a Righi oscillator placed at the focus of a parabolic copper reflector two or three feet in diameter. The waves sent out by this oscillator are quite different from the others, being only about two feet long, instead of three or four hundred feet, and the results, up to the present, are less important than those obtained with the pendent wire. Still in trials on the Salisbury Plain, he and his assistants sent messages perfectly in this way over a distance of a mile and three-quarters, and were able to direct these messages at will by aiming the reflector in one direction or another. It appears that these Hertzian waves, though invisible, may be concentrated by parabolic reflectors into parallel beams and projected in narrow lines, just as a bull's-eye lantern projects beams of light. And it was found that a very slight shifting of the reflector would stop the messages at the receiving end. In other words, unless the Hertzian beams fell directly on the receiver, there was an end of all communication.

"Do you think," I asked, "that you will be able to send these directed messages very much farther than you have sent them already?"

"I am sure we shall," said Marconi. "It is simply a matter of experiment and gradual improvement, as was the case with the undirected waves. It is likely, however, that a limit for directed messages will be set by the curvature of the earth. This stops the one kind, but not the other."

"And what will that limit be?"

"The same as for the heliograph, fifty or sixty miles."

"And for the undirected messages there is no limit?"

"Practically none. We can do a hundred miles already. That only requires a couple of high church steeples or office buildings, New York and Philadelphia, with their skyscraping structures, might talk to each other through the ether whenever they wished to try it. And that is only a beginning. My system allows messages to be sent from one moving train to another moving train or to a fixed point by the tracks; to be sent from one moving vessel to another vessel or to the shore, and from lighthouses or signal stations to vessels in fog or distress."

Marconi pointed out one notable case where his system of sending directed waves might render great service to humanity. Imagine a lighthouse or danger spot in the sea fitted with a transmitter and parabolic reflector, the whole kept turning on an axis and constantly throwing forth impulses in the ether—a series of danger signals, one might call them. It is evident that any vessel fitted with a Marconi receiver would get warning through the ether (say by the automatic ringing of a bell) long before her lookout could see a light or hear any bell or fog-horn. Furthermore, as each receiver gives warning only when its rotating reflector is in one particular position—that is, facing the transmitter—it is evident that the precise location of the alarm station would at once become known to the mariner. In other words, the vessel would immediately get her bearings, which is no small matter in a storm or fog.

Again, the case of lightships off shore gives the Marconi system admirable opportunity of replacing cables, which are very expensive and in constant danger of breaking. In December, 1898, the English lightship service authorized the establishment of wireless communication between the South Foreland lighthouse at Dover and the East Goodwin lightship, twelve miles distant; and several times already warnings of wrecks and vessels in distress have reached shore when, but for the Marconi signals, nothing of the danger would have been known. One morning in January, for instance, during a week of gales, Mr. Kemp, then stationed at the South Foreland lighthouse, was awakened at five o'clock by the receiver bell, and got word forthwith that a vessel was drifting on the deadly Goodwin Sands, firing rockets as she went. At this moment there was so dense a fog bank between the sands and the shore that the rockets could never have been seen by the coast-guards. They were now, however, informed of the crisis by telegraph, and were able to put out at once in their life-boats.

At another time, also in heavy fog, a warning gun sounded from the lightship, and at once the receiver ticked off: "Schooner headed for sands. Are trying to make her turn."

"Has she turned yet?" questioned Kemp.

"No. We've fired another gun."

"Has she turned yet?"

"Not yet. We're going to fire again. There, she turns." And the danger was over without calling on the life-boat men, who might otherwise have labored hours in the surf to save a vessel that needed no saving. Another application of wireless telegraphy that promises to become important is in the signaling of incoming and outgoing vessels. With Marconi stations all along the coast it would be possible, even as the discovery stands to-day, for all vessels within twenty-five miles of shore to make their presence known and to send or receive communications. So apparent are the advantages of such a system that in May, 1898, Lloyds began negotiations for the setting up of instruments at various Lloyds stations; and a preliminary trial was made between

Ballycastle and Rathlin Island in the north of Ireland. The distance signalled over here was seven and a half miles, with a high cliff intervening between the two positions; the results of many trials here were more than satisfactory.

I come now to that historic week at the end of March, 1899, when the system of wireless telegraphy was put to its most severe test in experiments across the English Channel between Dover and Boulogne. These were undertaken at the request of the French Government, which is considering a purchase of the rights to the invention in France. During the several days that the trials lasted, representatives of the French Government visited both stations, and observed in detail the operations of sending and receiving. Mr. Marconi himself and his chief engineer, Mr. Jameson Davis, explained how the installations had been set up and what they expected to accomplish.

At five o'clock on the afternoon of Monday, March 27th, everything being ready, Marconi pressed the sending-key for the first cross-channel message. There was nothing different in the transmission from the method grown familiar now through months at the Alum Bay and Poole stations. Transmitter and receiver were quite the same; and a seven-strand copper wire, well insulated and hung from the sprit of a mast 150 feet high, was used. The mast stood in the sand just at sea level, with no height of cliff or bank to give aid.

"Brripp — brripp — brripp — brripp—brrrrrr," went the transmitter under Marconi's hand. The sparks flashed, and a dozen eyes looked out anxiously upon the sea as it broke fiercely over Napoleon's old fort that rose abandoned in the foreground. Would the message carry all the way to England? Thirty-two miles seemed a long way.

"Brripp—brripp—brrrrr— brripp—brrrrr—brripp—brripp." So he went, deliberately, with a short message telling them over there that he was using a two-centimeter spark, and signing three V's at the end.

Then he stopped, and the room was silent, with a straining of ears for some sound from the receiver. A moment's pause, and then it came briskly, the usual clicking of dots and dashes as the tape rolled off its message. And there it was, short and

The royal yacht "Osborne" from which the Prince of Wales telegraphed without wires. The sending and receiving wire is suspended from the rope connecting the two mastheads, and can be distinguished by the wire cone near the top. From a photograph by A. E. Beken.

commonplace enough, yet vastly important, since it was the first wireless message sent from England to the Continent: First "V," the call; then "M," meaning, "Your message is perfect;" then, "Same here 2 c m s. V V V," the last being an abbreviation for two centimeters and the conventional finishing signal.

And so, without more ado, the thing was done. The Frenchmen might stare and chatter as they pleased, here was something come into the world to stay. A pronounced success surely, and everybody said so as messages went back and forth, scores of messages, during the following hours and days, and all correct.

On Wednesday, Mr. Robert McClure and I, by the kindness of Mr. Marconi, were allowed to hold cross-channel conversation, and, in the interests of our readers, satisfy ourselves that this wireless telegraphy marvel had really been accomplished. It was about three o'clock when I reached the Boulogne station (this was really at the little town of Wimereux, about three miles out of Boulogne). Mr. Kemp called up the other side thus: "Moffett arrived. Wishes to send message. Is McClure ready?"

Immediately the receiver clicked off: "Yes, stand by;" which meant that we must wait for the French officials to talk, since they had the right of way. And talk they did, for a good two hours, keeping the sparks flying and the ether agitated with their messages and inquiries. At last, about five o'clock, I was cheered by this service along the tape: "If Moffett is there, tell him McClure is ready." And straightway I handed Mr. Kemp a simple cipher message which I had prepared to test the accuracy of transmission. It ran thus:

McClure, Dover : Gniteerg morf Ecnarf ot Dnalgne hguorht eht rehte.

Moffett

Read on the printed page it is easy to see that this is merely, "Greeting from France to England through the ether," each word being spelled backward. For the receiving operator at Dover, however, it was as hopeless a tangle of letters as could have been desired. Therefore was I well pleased when the Boulogne receiver clicked me back the following:

Moffett, Boulogne : Your message received. It reads all right.
Vive Marconi.

McClure

Then I sent this:

> Marconi, Dover : Hearty congratulations on success of first experiment in sending aërial messages across the English channel. Also best thanks on behalf of editors McClure's Magazine for assistance in preparation of article.
>
> *Moffett*

And got this reply:

> Moffett, Boulogne : The accurate transmission of your messages is absolutely convincing. Good-by.
>
> *McClure*

Then we clicked back "Good-by," and the trial was over. We were satisfied: yes, more, we were delighted.

I asked one of Marconi's chief engineers if the Boulogne and Dover installation would remain permanent now. He said that depended on the French and English governments. The latter has a monopoly in England on any system of telegraphy in which electric apparatus is used; and all cross-channel cables are of British ownership.

"There must be a great saving by the wireless system over cables," I said.

"Judge for yourself. Every mile of deep-sea cable costs about $750; every mile for the land-ends about $1,000. All that we save, also the great expense of keeping a cable steamer constantly in commission making repairs and laying new lengths. All we need is a couple of masts and a little wire. The wear and tear is practically nothing. The cost of running, simply for home batteries and operators' keep."

"How fast can you transmit messages?"

"Just now at the rate of about fifteen words a minute; but we shall do better than that no doubt with experience. You have seen how clear our tape reads. Any one who knows the Morse code will see that the letters are perfect."

"Do you think there is much field for the Marconi system in overland transmission?"

"In certain cases, yes. For instance, where you can't get the right of way to put up wires and poles. What is a disobliging farmer going to do if you send messages right through his farm, barns and all? Then see the advantage in time of war for quick communication, and no chance that the enemy may cut your wires."

"But they may read your messages."

"That is not so sure, for besides the possibility of directing the waves with reflectors, Marconi is now engaged in most promising experiments in syntony, which I may describe as the electrical tuning of a particular transmitter to a particular receiver, so that the latter will respond to the former and no other, while the former will influence the latter and no other. That, of course, is a possibility in the future, but it bids fair soon to be realized. There are even some who maintain that there may be produced as many separate sets of transmitters and receivers capable of working only together as there are separate sets of Yale locks and keys. In that event, any two private individuals might communicate freely without fear of being understood by others. There are possibilities here, granting a limitless number of distinct tunings for transmitter and receiver, that threaten our whole telephone system. I may add, our whole newspaper system."

"Our newspaper system?"

"Certainly; the news might be ticked off tapes every hour right into the houses of all subscribers who had receiving-instruments tuned to a certain transmitter at the news-distributing station. Then the subscribers would have merely to glance over their tapes to learn what was happening in the world."

We talked after this of other possibilities in wireless telegraphy and of the services Marconi's invention may render in coming wars.

"If you care to stray a little into the realm of speculation," said the engineer, "I will point out a rather sensational role that our instruments might play in military strategy. Suppose, for instance, you Americans were at war with Spain, and wished to keep close guard over Havana harbor without sending your fleet there. The thing might be done with a single fast cruiser in this way: Supposing a telegraphic cable laid from Key West, or any convenient point on your shores, and ending at the bottom of the sea a few miles out from the harbor. Let us imagine this to have been done without knowledge of the Spaniards. And suppose a Marconi receiving-instrument, properly protected, to be lying there at the bottom in connection with the cable. Now, it is plain that this receiver will be influenced in the usual way by a Marconi transmitter aboard the cruiser, for the Hertzian waves pass well enough through water. In other words, you can now set the armature of a relay down at the ocean's bottom clicking off Morse signals as fast as you like, and it is a simple matter of electrical adjustment to make that armature repeat these signals automatically over the whole length of cable in the ordinary way.

"With this arrangement, the captain of your cruiser may now converse freely with the admiral of the fleet at Key West or with the President himself at Washington, without so much as quitting his deck. He may report every movement of the Spanish warships as they take place, even while he is following them or being pursued by them. So long as he keeps within twenty or thirty miles of the submerged cable-end, he may continue his communications, may tell of arrivals and departures, of sorties, of loading transports, of filling bunkers with coal, and a hundred other details of practical warfare. In short, this captain and his innocent-looking cruiser may become a never-closing eye for the distant American fleet, an eye fixed continually upon an enemy all unsuspicious of this communication and surveillance. And it needs but little thought to see how easily an enemy at such disadvantage may be taken unawares or be led into betraying important plans."

This conception struck me as so interesting that I pressed my informant to say how far he thought it lay in the realm of speculation.

"Why," said he, "it is a sensible enough little dream that might be realized, if any one cared to spend the money and take the necessary trouble. There is no doubt our instruments could be made to operate a cable at sea-bottom, just as they could be made to blow up a powder magazine in a beleaguered city or steer a ship from a distance, or"

"Steer a ship from a distance?" I interrupted.

"Certainly, a small one, say a lightship, with no one aboard her."

"How could you steer her?"

"Oh, by a simple arrangement of commutators and relays. It isn't worth while going into the thing, but you could send one signal through the ether that would part her cables, say by an explosive tube or a simple fusing process. Then you could send another signal that would open her throttle-valve and start her engines. Of course, I'm assuming fires up and boilers full. Then you could send other signals that would put her helm to starboard or port and so on. And straightway your lightship would go where you wanted her to. There may not seem to be much sense in steering an empty lightship about, but don't you see the vast usefulness in warfare of such control over certain other craft? Think a moment."

He smiled mysteriously while I thought.

"You mean torpedo craft?"

"Exactly. The warfare of the future will have startling things in it; perhaps the steering of torpedo craft from a distance will be counted in the number. But we may leave the details to those who will work them out."

41

And here, I think, we may leave this whole fascinating subject, in the hope that we have seen clearly what already is, and with a half discernment what is yet to be.

The Great Dynamite Factory at Ardeer

The Making and Handling of High Explosives.—Life and Manners of the Workmen.—Precautions Against Accidents.—The Small Number of Casualties.

By H. J. W. Dam

August 1897

THE great dynamite factory at Ardeer in Scotland, the largest of its kind, is one of the most picturesque places in the world. Considering the unique and dramatic conditions that prevail among its workers, the neglect of Ardeer hitherto by novelists and dramatists is surprising. This may be due, however, to the fact that it is exceedingly difficult for a stranger to obtain access to the factory, while, once inside, the surroundings are rather trying to sensitive nerves. For six hours a day and two days in succession your life depends, at every moment, upon a thermometer.

Great is the thermometer at Ardeer! Nitroglycerin, a teaspoonful of which would blow you to fragments, surrounds you in hundreds and thousands of gallons. It is making itself in huge tanks, gurgling merrily along open leaden gutters, falling ten feet in brown waterfalls, so to speak, into tanks of soda solution, and bubbling so furiously in other cylinders, through the in-rush of cold air from below, that it seems to be boiling. It is being drawn off from large porcelain taps like ale, poured into boxes, and rattled along tramways. In the form of dynamite, it is being rubbed with great force through brass sieves, jammed into cartridges, and flung into boxes; and in the form of blasting gelatin, it is being torn by metal rods, forced through sausage machines, and cut, wrapped, and tossed into hoppers—all these processes proceeding as rapidly as if it were ordinary olive-oil instead of the deadliest explosive known to man.

All around you are big cotton mills and storehouses as full of fleecy, white cotton as ordinary cotton mills and storehouses, but every pinch of the cotton, still white and fleecy, has been nitrated into gun-cotton, and would suffice, if exploded, to cut you off in the beauty of your youth. Death, instantaneous and pulverizing, encircles you, in fact, by the ton; but the man and the thermometer surround you also. The man's eyes never leave the instrument. Both are chosen for their perfect reliability; and endless precautions, innumerable rules, and the strictest discipline maintain Ardeer in a state of busy and peaceful security, and prevent it from being scattered periodically over the calm blue sea that widens endlessly on one side, or the hungry brown acres of Scotland which stretch away to the horizon on the other.

THE NITROGLYCERIN "HILLS."

From the top of one of the nitroglycerin "hills" the factory looks like an enormous and eccentric landscape garden. In every direction rise green embankments, square, conical, or diamond-shaped, from fourteen to seventy feet in height, and covered with long rank grass. Many of them are faced with corrugated iron, and look like high fences. From the top of each mound peeps the red canvas roof of a white wooden house—a house within a hill—which is from one to four stories in height. Every explosive structure is surrounded by artificial banks, so that in the event of an accident all the others will be protected from concussion or flying fragments. There are three nitroglycerin "hills"; and on the one before you the nitrating-houses, two in number, in which the nitroglycerin is made, stand out in clear relief at the top. They are frail wooden cabins, which were expected by Mr. Nobel when he built them to last six months, but which have not yet been blown to pieces after twenty-five years of constant use. Tunnels through the banks open everywhere. Tramways and lines of pipes on trestles cross each other diversely. This is the "Danger Area," the wide expanse in which the explosives are made and moved about. It is surrounded in an irregular semicircle by fourteen large groups of structures, from which rise fourteen high chimney-stacks. These include the nitric-acid works, acid recovery, ammonia-mill, potash-mill, "guhr"-mill, steam and power houses, box-factories, washing, carding, and bleaching departments for the cotton, pulping-mills, and other contributing industries, connected by steam railway tracks which join the Glasgow line. There are 450 separate structures, now occupying 400 acres out of the 600 owned by the company, which were, when the site was chosen by Mr. Nobel in 1871, a barren waste of sand dunes, stretching for a mile

A nitroglycerin "hill" at Ardeer.
The nitroglycerin is made in the two houses at the top of the hill, and washed in those immediately beneath. The house in the center is a "drowning-tank," and that at the bottom of the hill is the "final" washing-house. "Every explosive structure is surrounded by artificial banks, so that in the event of an accident all the others will be protected from concussion or flying fragments."

and three-quarters along the sea.

Into this kingdom of high explosives you enter by the courtesy of Mr. C. O. Lundholm, the works manager, under the guidance of the engineer of the works, Mr. E. W. Findlay. The strain upon your nerves begins mildly. Your hair is quite ready to rise, so ready that you can feel it awake and stretch itself at every spot of grease—which may be nitroglycerin—and every stray pinch of cotton—which may be gun-cotton. You now understand for the first time the psychological condition of a shying horse. You go along just as the horse does, with eyes strained at every small object and a lurking predis-position to bolt.

The acid-works are soothing, however. They are quite safe. Nitroglycerin is made from glycerin, the sweetish adjunct of the dressing-table, and nitric acid. The glycerin is bought by hundreds of tons from various sources. In this big barn which you enter the nitric acid is manufactured. In two rows stand fifty-eight steel retorts about six feet in diameter and four feet deep, which are bricked up like ovens. Here sulphuric acid, or oil of vitriol, from Glasgow is combined with nitrate of soda from Chili, and the nitric acid thus set free passes over in pipes to a high framework carrying numberless brown earthenware jars in which it condenses. As it passes over it gives off reddish fumes which are suffocating—a whiff of them gives you a fit of coughing, and a full breath of them would choke a locomotive. Mr. Findlay explains that the nitric acid thus made is mixed with a larger quantity of sulphuric acid, and moved in steel pony-cars to a station at the foot of each nitroglycerin "hill." Thence the acids are drawn up by cable or blown up through pipes to a tank at the top of the "hill" by compressed air. You mentally compare the advantages of being blown up with compressed air to being blown up by other means, and smoothing down your hair, enter the "Danger Area."

THE "DANGER AREA."

To enter the "Danger Area" you must pass the "searcher." He stands in front of his cabin, and you will find one of him always blocking the way at the four entrances to the explosive district. He is a tall, military-looking man in a blue uniform faced with red, and he takes from you all metallic objects—your watch, money, penknife, scarf-pin, match-case, matches, and keys. None of these are allowed to be where nitroglycerin is. He searches every man who enters, no matter how often the man may come and go. The girls, 200 of whom are employed, are not permitted to wear pins, hair-pins, shoe-buttons, or metal pegs in their shoes, or carry knitting, crochet, or other needles. These regulations are the outgrowth of experience and the long-ago discovery in dynamite cartridges of buttons and other foreign substances calculated to make trouble at unexpected moments. The girls are searched thrice a day by the three matrons who have them in charge. From the lack of hair-pins they wear their hair in braids, tied with ribbons, which gives them all an unduly youthful look. The searcher tells you that his chief trouble is with matches. Some of the lower-class male employeese—there are

The "searcher" at work at the entrance to a nitroglycerin hill.

"He searches every man who enters, no matter how often the man may come and go."

1,100 men in the factory—are willing at times to smuggle in matches for a quiet smoke in a secluded corner. This quiet smoke may of course produce a much louder smoke in a corner not secluded, and is therefore rigidly banned. The discipline in the factory is most extraordinary, and to it must be attributed the marvelous immunity from accidents.

At this point, too, you get your first glimpse of the "costumes." A man in a Tam o' Shanter cap comes up clothed from head to foot in vivid scarlet. He belongs to a nitroglycerin house. Then comes a man in dark blue, a "runner" or carrier of explosives. Then comes a man in light blue, who belongs to a smokeless-powder factory. All the girls are in dark blue. The different colors are used so that a superintendent at any distance can always tell if a man is on his own ground and attending to his own work. A few weeks since, a cartridge lassie in dark blue said to a man in scarlet, "Gi'e us a kiss," and he promptly "gi'ed" her one. This unlawful combination of colors caught the eye of an overseer hundreds of yards away, and the pair were instantly removed from the works and the pay-roll. Kissing and skylarking are absolutely prohibited during working hours, but on Saturdays and Sundays the workers make full amends. If reports are to be believed, the workers are more than usually romantic in their tendencies, the alleged cause being the constant breathing of nitroglycerin; and inquiring Pickwicks have taken many notes thereupon, in which the statistics of marriage and population are not entirely neglected.

THE NITRATING-HOUSES.

Having passed the searcher, you mount the "hill," an artificial one, built of sand, and perhaps sixty feet high. On the top of it are two "nitrating-houses." They are of thin clapboards painted white, and are about twenty feet square. These houses are always placed on the tops of "hills," in order that the nitroglycerin, passing from process to process, may flow by its own weight downward. It is not exactly the kind of liquid that one wants to pump. At the door of the house you are confronted by two pairs of yawning rubber shoes. Large shoes of rubber, indeed, and sometimes

A dynamite cartridge house.

In the small cabin before the house the girls stop to remove their walking shoes before going to their work.

even larger ones of leather confront you at the door of every danger house. No shoe which touches the ground outside is allowed to touch the floor of a danger department. The least grit might make friction and lead to an explosion. In all departments the girls are compelled to change to slippers or work barefooted, the majority, in summer, preferring the latter. Having stepped into the overshoes, you begin to flop like a great auk over the sheet-lead which covers the floor. The shoes are trying, particularly as you have other things to worry you. Snow-shoes, ski, and stilts can all be practiced on with advantage before endeavoring to get about in a pair of overshoes which do not fit your own shoes and are ceaselessly trying to trip you up.

As you enter the nitrating-house your eye is caught by two lead cylinders, five feet in diameter and six feet deep, which are sunk in the floor. They have closed, dome-shaped tops, over which many lead pipes curl and into which they enter!

At the farther cylinder sits a man in scarlet watching a thermometer. He neither moves, looks up, nor betrays any sign of your presence. The thermometer which he is watching is five feet in length. Only the top or marked portion extends above the cylinder, the tube which carries the mercury reaching down to the hot acids and nitroglycerin. In the cylinder has been placed about

Interior of a mixing-house.

Here the mixture of kielselguhr, carbonate of ammonia, and nitroglycerin, which makes dynamite, is thoroughly worked by hand and put through a sieve.

a ton and a half of sulphuric acid mixed with a ton of nitric. Into this mixture are now being sprayed 700 pounds of glycerin, the glycerin injector-pipe being joined by another carrying compressed air. As fast as the glycerin spray enters the mixture it seizes the nitrogen of the nitric acid and combines to nitroglycerin, and the sulphuric takes up the water which is thus set free. The process requires fifty-five minutes, during which the 700 pounds of glycerin becomes about 1,500 of nitroglycerin. Great heat is caused by the chemical action, and the absolute necessity is that the heat shall be kept down or it will explode the newly formed nitroglycerin. To this end the cylinder is surrounded by a water-jacket, through which cold water is rushing constantly, and four concentric coils of lead pipe occupy the interior of the cylinder, carrying four steady rushes of cold water.

If the heat, through vagaries in the glycerin, rose above the danger point, the thermometer would instantly reveal this to the man on watch. If the thermometer rose ever so little above twenty-two degrees centigrade, the man would turn on more air and shut off the inflow of glycerin. If it continued to rise slowly and he could not stop it by more air and water, he would give a warning shout, "Stand by," to a man watching below. If it continued, he would shout "Let her go," and the man would open a valve; this would sweep the whole charge down to a "drowning-tank" lower down the hill, which would drown the coming explosion in excess of water. The two men the meanwhile would bolt to a safe position behind banks. If the heat rose rapidly, too rapidly for "drowning," the man would pull the valve, give a warning shout, and run. So would everybody, you included. You might run on one side to the protecting arms of a dynamite magazine holding twenty tons, or on the other to the soothing shelter of a house where gun-cotton is baking at 120 degrees Fahrenheit. Failing these, there is the pond. This is a sweet, placid pond which is formally blown up once a week because some dregs of nitroglycerin have drained into it and collected at the bottom, making it unsafe. It is comforting to feel, in the hour of danger, that you have havens of perfect security such as these.

The glycerin having duly become nitroglycerin, you flop down the stairs to another department, to witness its separation from the acids with which it is now mixed. It comes shooting down a lead gutter, and falls, a cream-colored stream, to the bottom of a lead tank, eight feet in length and two in width. As soon as the tank is full, the nitroglycerin, lighter than the acid, rises to the surface like oil. It is skimmed off in an aluminium skimmer resembling a tin wash-hand basin with a handle, and is poured into a lead pocket at the end, whence it flows through pipes to a tank, where it receives

Making blasting gelatin cartridges

"Blasting gelatin, a yellow, tough elastic paste, …is being forced through a sausage machine, and chopped, by hand, into three-inch lengths."

its first washing with cold water. Thence it goes through gutters farther down to another department, where it is washed with warm water and carbonate of soda. Every particle of the free acid must be removed, as remnants of it might cause chemical action, heat, and explosion in the dynamite or blasting gelatin later on. A sample is taken of each lot of nitroglycerin when made. This is placed in a small clear glass bottle and covered with blue litmus solution, to detect the presence of any remaining free acid, which would color the litmus red. *En passant*, your guide mentions that some years ago one of the foremen was carrying a little felt-lined box of these samples to one of the sample magazines when he unfortunately stumbled and fell. He was blown to pieces.

You have now reached the bottom of the "hill" (all nitroglycerin factories are called "hills"), and are in a wooden cabin, with a floor of loose sand, where the making of dynamite and blasting gelatin actually begins. Dynamite consists merely of liquid nitroglycerin which has been absorbed by some

Making dynamite cartridges

"The girls work with the greatest rapidity ... The sliding brass rod of the machine is actually lubricated with nitroglycerin."

porous material. The liquid was discovered by Sobrero, an Italian, in 1846. Its transport and use were attended with such danger, however, that the late Alfred Nobel conceived, in 1867, the plan of absorbing it in some non-explosive medium. After experimenting with saw-dust, brick-dust, charcoal, paper, rags, and kieselguhr, he finally settled upon the last named as the best material. Kieselguhr, known in the factory as "guhr," is a silicious earth, mainly composed of the skeletons of mosses and microscopic diatoms, which is found as a slaty black peat in Scotland, Germany, and Italy. Before being used it goes to the "guhr-mill," where it is calcined in a large kiln, rolled, and sifted, the result being a very light pink powder of the consistency of flour. In the house you have entered, twenty-five pounds of kieselguhr, with about one pound of carbonate of ammonia, are weighed into a wooden box about three feet square and eighteen inches deep. Upon it is drawn seventy-five pounds of nitroglycerin from the filter tank by a man in scarlet. Another man in scarlet, with his arms bare to the shoulders, takes the box to a table, and gives it a preliminary mix, to see that all the nitroglycerin is roughly absorbed. Then a man in blue seizes it, places it with other boxes on his hand-car or "bogie," and pushes the load off to the "mixing-houses."

A DISASTROUS EXPLOSION—THE MIXING-HOUSES.

At half-past six on the morning of the 24th of February, one week after the writer's visit to this house, it was the scene of a very disastrous explosion. Twenty-four hundred pounds of nitroglycerin was collected here, in the tanks and boxes mentioned, and from some cause which may never be known it exploded, killing six people—a chemist, a foreman, and four workmen. A few other employees were slightly hurt by flying débris. The sound was of course tremendous, and the effects of the explosion, which were very, clear at Irvine, three and one-half miles away, are said to have been so strong in a town ten miles away that the gas-lamps were extinguished by the air concussion. A disaster such as this, whose suddenness is not its least painful characteristic, cannot of course be minimized in its tragic importance. At the same time, it serves as the best possible testimony to the value of the system of protection employed. That over a ton of nitroglycerin can explode in the heart of a factory where 1,300 people are at work, and only the six men, within a few feet of it, lose their lives, shows better than any other evidence the meaning and value of the Ardeer mounds.

You follow the box to a "mixing-house." This, in the case of dynamite, is a large wooden cabin, containing a long narrow table on each side. In it six girls are at work. The runner sets the open box of the mixture down in the doorway. A girl hoists it to a ta-ble, and flies at it with bare arms as if it contained only flour and water. She mixes it thoroughly. Then she takes a big wooden scoop, jabs it into the box, and dumps the scoopful into a raised box of the same size, with a brass sieve bottom. She then, as if the sieve bottom were a washing-board, rubs the dynamite with all her strength against the sieve, forcing it through the small holes. A few of the girls use a leather hand-flap to rub

Reading the thermometer before entering the testing magazine "India."

It is in "India" that the company's explosives are tested through long periods under high heat and severe cold.

with, but most of them prefer their bare hands. You view the process with consternation. Hitherto you have looked upon dynamite as something to be regarded politely from a safe distance as if it were a rattle-snake. The girls handle it, however, as coolly as if it were the sand on the floor. Some of it is continually spilt, of course, and mixes with this sand, but the sand is all re-moved at short intervals and buried. One of the few fatal accidents in the his-tory of Ardeer took place near this house. A cartridge hut wherein four girls were working exploded, killing the girls. Burning dust from this hut fell into the open boxes of dynamite in three other huts. The dynamite began to blaze, and the deadly smoke from it, which consists of hyponitric-acid fumes, imme-diately filled the huts. Two girls in each hut had the courage to jump over the blazing boxes, and escaped; but the others, six in number, were suffocated in a few minutes. Thus, ten persons lost their lives. When the huts were en-tered, the six girls were found seated in perfectly natural attitudes, their fac-es showing no trace of agony or fear. It was evident that, having been stunned by the sudden explosion, they had been suffocated before recovering from the shock. It will be noted that the loose dynamite burned and did not explode. This is one of several curious facts concerning dynamite which will be considered later.

It maybe well to state at this point that the two hundred and odd young ladies employed in this dangerous work are all strictly beautiful. Everybody who visits the factory admits this at once. Nobody, in fact, seems inclined to invidious comparisons among strong and courageous girls, when each of them has enough dynamite in her possession to blow a hole in Scotland. Moreover, there is some reason for the statement. The breathing of nitroglycerin by the workers gives them a universal clearness of skin, and among the fairer girls the contrast of scarlet and white in their faces is most unusual. You learn that (perhaps in consequence of their complexions) the girls marry quickly after entering the factory.

THE CARTRIDGE HOUSES.

After being rubbed through the sieves the dynamite becomes a finely divid-ed, greasy, coffee-colored earth. It is now the dynamite of commerce, and is ready to be made into cartridges. As you approach one of the cartridge houses, which are small white one-story buildings, you hear a tremendous thumping. You ask your guide in some perturbation if it is a good day to look at cartridge houses, but he smiles and says that the noise is merely the cartridge machines. The hut is about ten feet square, with a single door. Four girls are at work. Against the right and left walls are four spring pump-handles about the height of a girl's head. Each pump-handle when pulled down forces a brass rod through a small conical hopper of loose dynamite fixed to the wall, and jams a portion of the dynamite down a brass tube at the bottom of the box. The girl wraps a small square of branded parchment paper around the bottom of the tube, folding it at the lower end. Then, holding the paper with one hand, and jumping up and down as she works the pump-handle with the other, she push-es dynamite down the tube till the paper cylinder is filled to a depth of about three inches. She then removes it, folds down the top of it, drops it through a slide in the wall, whence it rolls down into her own special box a finished car-tridge. She replenishes her stock of dynamite with a scoop through a sliding door in the wall, from a box of loose dynamite which the runner has placed in a closed chest immediately outside. The girls work with the greatest rapidity. The sliding brass rod is actually lubricated with nitroglycerin. To see this opera-tion—the brass rods flying up and down, damp with nitroglycerin, and dyna-mite being forcibly jammed down a brass tube—entirely destroys your appetite for further knowledge. It is incredible, and you want to go away, outside the "Danger Area," and think it over. But your guide takes you instead to a blasting gelatin cartridge hut. Here blasting gelatin, a yellow, tough, elastic paste, which consists of about seven per cent, of nitro-cotton and ninety-three of nitro-glycerin, is being forced through a sausage machine, chopped, by hand, into three-inch lengths with a wooden wedge upon a lead-covered table, and wrapped into cartridges, at the greatest speed. Blasting gelatin is fifty per cent, more powerful than dynamite, and the effect on your mind is to make you exactly fifty per cent, more uncomfortable than before; to multiply by one and one-half your desire to get away before any *contretemps* occurs which you would be in no position to either explain or avoid.

There are forty-five cartridge huts, all heated by steam to not less than fifty degrees Fahrenheit. Nitroglycerin congeals at forty-three Fahrenheit and freezes at forty, so the huts must be kept warm. If the dynamite were allowed to rest against a steam-pipe an explosion might follow, and the pipes are carefully boxed, and the thermometer is always watched by the eye of authority. In addition to dynamite and blasting gelatin cartridges, the company, manufacture cartridges of gelatine dynamite and gelignite, combinations of nitroglycerin, nitro-cotton, nitrate of potash, and wood meal. The gelatin explosives are specially adapted for use under water, being entirely unaffected by damp-

Interior of the barn-like building where nitro-cotton is made.

To make nitro-cotton, cotton waste is mixed nith sulphuric and nitric acid.... "In a few minutes the chemical combination takes place, the acid is poured off, and the nitro cotton receives its first washing."

ness of any kind. The company also make "Ardeer powder" and "carbonite"—explosives for blasting purposes in fiery coal mines, with a lower percentage of nitroglycerin than dynamite. The output of explosives of all kinds is an average of about one hundred tons per week.

MAKING NITRO-COTTON ON A MAMMOTH SCALE.

Nitro-cotton, which by itself and in combination with nitroglycerin as cordite and ballistite is rapidly displacing gunpowder in every direction, is made and used by the ton at Ardeer. It is made from cotton-waste, the waste left on the spindles in the cotton-mills. This comes to Ardeer in bales, like bales of finished cotton, and is first washed, to remove all grease and dirt, carded, and reduced to a homogeneous mass in a big mill devoted to these processes. Then it goes to a great barn-like building where it is turned into soluble nitro-cotton or insoluble gun-cotton, as may be desired, the process taking place in small iron pans or hundreds of earthenware jars. Half the floor is taken up by these jars, which sit side by side in a shallow tank of cement about a foot deep. The object of this tank is to keep the jars cool by surrounding them with water during the nitration. Along one side of the room are the acid taps and lead pans. Four pounds of cotton are placed in a pan, and one hundred and fifteen pounds of mixed sulphuric and nitric acid are added. In a few minutes the chemical combination takes place, the acid is poured off, and the nitro-cotton receives its first washing. From this point, until every particle of the acid has been washed out of it, it is liable to burn spontaneously at any instant. As one of the workmen dumps the pan load into the "centrifugal" or acid separator, it may go up with a flash and a great column of yellow smoke; and this not unfrequently happens, but does no great harm except, perhaps, to beards and eyebrows. It takes fire slowly and gives full warning. It now goes to another department and is washed repeatedly, kept for a week in water tanks, pulped in ordinary pulping-mills, and dried in rotary centrifugal machines until all but thirty per cent, of the water is eliminated. The remainder is dried out of it on the shelves of a great drying-house, where a temperature of from 100 to 120 degrees Fahrenheit is maintained by hot air through fans.

At Ardeer this nitro-cotton is used in enormous quantities in combination with nitroglycerin to make blasting gelatin, of which it contributes seven per cent.; cordite, of which it is forty per cent.; and ballistite, which consists of sixty per cent, of soluble nitro-cotton and forty per cent, nitroglycerin. The extraordinary affinity of soluble nitro-cotton for nitroglycerin is a curious chemical fact. No matter how much water is present in the mixing-tank, every par-

The man and thermometer in one of the nitrating-houses

"Death, instantaneous and pulverizing, encircles you, in fact, by the ton: but the man and the thermometer surround you also. The man's eyes never leave the instrument."

ticle of gun-cotton will find and absorb the nitroglycerin, and this "wet-mixing process" as invented and carried on at Ardeer is admirable of its kind. The material for cordite, in the form of cordite paste, is made in large quantities at Ardeer, and sent to the government factory at Waltham, where the government smokeless ammunition is made. Ballistite is a specialty at Ardeer, and is rapidly displacing the other smokeless powders for sporting purposes. Its admirers claim that it is stronger than any other, cleaner in the gun, perfectly smokeless, and entirely unaffected by heat or dampness. It can be soaked in water and fired without any loss of efficiency. Since the professional pigeon shots have largely adopted it, and the weekly scores in the sporting papers show the majority of kills to its credit, the shot-gun fraternity, so numerous in England, have taken to it *en masse*. Ballistite is made in three forms: in cubes for cannon, in minute rings for rifles, and in square flakes for shot-guns. As first made and dried, it is a light brown, elastic paste. This is run through steel rollers which are heated to 120 degrees till it becomes as thin as tissue paper and transparent. It is like thin, elastic sheets of silky horn. Then it is cut up in cutting-machines into grains of various sizes for rifles or shot-guns, as the case may be.

These processes are most ingenious and mechanically interesting, and occupy several large mills by themselves. In all are the thermometers and the shoes. The machinery in nearly all cases represents original inventions, either conceived in Ardeer or invented by Mr. Nobel, who was the originator of smokeless powders. Absolute cleanliness reigns. Dust is never allowed to collect, and the small quantity of sweepings from the leaden floors are daily burned.

The subsidiary departments are full of interest. "India" and "Siberia" are two magazines where the company's explosives and others from all sources are tested through long periods under high heat and severe cold respectively. "India" is of course the more dangerous, and before entering it your guide climbs a ladder on the embankment which surrounds it and peeps through a three-inch hole to read the thermometer projecting from the roof of the house inside. "India" caught fire in 1895, and would have harmed nothing but itself had not some over-eager firemen gone inside the banks and attempted to extinguish the fire. In the explosion which occurred two were killed and two other employees injured. To avoid a repetition of this occurrence a huge sprinkler now rises in the center of the hut, by means of which at the first sign of fire the whole interior can be deluged from a safe distance. A thermo-electric "tell-tale" also runs from "India" to a laboratory.

In the packing-houses the cartridges are packed by girls into five-pound cardboard boxes, which in turn are grouped in fifty-pound wooden cases. These cases are taken in hand-cars to the magazines and thence to the beach, the railways running into the sea. The cases are transferred to boats and loaded into the company's own steamers, which carry them to all the Channel and neighboring ports for shipment all over the world. There are also sample magazines, an armory containing all the ancient and modern small arms; a shooting range, with its attendant officers and experts, where the explosives for rifles and shot-guns are carefully tested; laboratories, and contributing departments of all kinds.

REMARKABLE FREEDOM FROM CASUALTIES.

Having now inspected the factory in all its interesting entirety, you are confronted with a statement so extraordinary as to be almost incredible, viz., that despite the manufacture by the ton of all these deadly explosives, Ardeer is one of the safest factories that you could possibly be in. In the whole period of its existence, about twenty-five years, the entire loss of life by accidents, including the sad occurrence of February 24th, has been only twenty-one. This, compared with the number of people employed, is lower than the death-rate in any cotton-mill, woolen-mill, foundry, boiler-shop, shipyard, or other large manufactory. The main cause of this excellent showing is the admirable character of the discipline imposed and the firm and careful system of management. But the rigid, intelligent, and systematic way in which explosive factories are guarded by government regulations and government inspectors undoubtedly also plays a large part in this result.

The nitroglycerin compounds, however, are far from being as dangerous as is generally supposed. Nitroglycerin itself is always a possible source of explosion, but up to this year no accident had ever attended its manufacture at Ardeer. The accidents that have occurred have been due to the handling of it after it has been made. With regard to dynamite, its actual safety as an explosive was ever the pride of its late inventor, Mr. Nobel. He claimed that dynamite could not be exploded by being thrown to the ground from any height; that it could sustain any degree of shock without explosion. He claimed for blasting gelatin that, in addition to being the strongest, it was absolutely the safest explosive known. In proof of this he devised a series of experiments which have been often performed at the factory and which have never failed. They may be seen at any time by a visitor whom the company desires to convince, and as given on a late occasion were as follows:

1. A cube of iron weighing 420 pounds was hoisted on crossed poles

above an ordinary packing-box containing fifty pounds of dyna-mite cartridges, the box resting on a board on the ground. The rope was cut by electrically exploding a cartridge against it, and the weight fell twenty-five feet, smashing the box completely and pulverizing some of the cartridges; but there was no explosion.

2. The same experiment was repeated with a box of blasting gel-atin cartridges, the fall being twenty-five feet and the iron weight 470 pounds. Box and contents were crushed and scat-tered, but there was no explosion.

3. A one-pound tin of gunpowder was placed on an open five-pound box of dynamite cartridges and exploded. The dynamite caught fire and burned up, but did not explode.

4. The same experiment was performed with a five-pound box of blasting gelatin cartridges with the same result.

5. A dynamite cartridge was set on fire by a fuse, and burned rather rapidly. It would have burned away completely, but a detonator had been placed in the middle, and when the flame reached this the other half of the cartridge exploded.

6. To show the strictly local force of dynamite, a one-pound car-tridge was hung eight inches above a three-eighths of an inch boiler-plate, which was lying on two bits of wood, and exploded. The plate was only slightly bent.

7. A similar cartridge was laid flat upon the same plate and ex-ploded, the result being a hole torn in the plate about the size of the cartridge.

8. A similar cartridge was then placed on a similar plate and covered with sand. Upon exploding, it tore a large hole in the plate.

Dynamite and blasting gelatin when set on fire will merely burn. If the dynamite is in a loose form, it will entirely burn away without danger. If com-pressed, both will burn until the heat reaches a point high enough to explode the remainder, but this always requires sufficient time to give bystanders full warning and enable them to reach a point of safety. All the nitroglycerin compounds are exploded by detonation; that is, by means of explosive caps like percussion caps which fit on the ends of the fuses. The cap explosive is a mixture of fulminate of mercury and chlorate of potash, and the Nobel company have a large and separate factory in Scotland which is devoted to the manufacture of fulminate of mercury and various kinds of detonators.

Shipping at Ardeer

The high explosives (dynamite and their cartridges in fifty-pound cases) are run into the sea on hand cars, lifted into boats, and finally put on board the company's steamers for shipment all over the world.

The explosive force of No. 1 dynamite, weight for weight, is four times that of gunpowder. Bulk for bulk, the dynamite being much heavier, it is over seven times as powerful as gunpowder. Blasting gelatin has nearly six times, weight for weight, and a fraction less than ten times, bulk for bulk, the power of gunpowder. Gun-cotton and No. 1 dynamite are about equal in explosive strength. Dynamite is not allowed on passenger trains in England, but is transported with great freedom on the Continent, and thirty thousand tons of it have been shipped on the English and Continental railways without accident up to date. Of course, every package and case carry explicit instructions, but that the danger is small the immunity from explosions in transport clearly shows.

The moral of which is, that dynamite is safe and blasting gelatin is safer if they are treated with only reasonable care. "The accidents do not occur here but in the use of it," says Mr. Johnston. "If the company's explicit printed instructions were followed, accidents would scarcely be known." Accidents often occur in thawing after an explosive has been frozen; but these arise from the incredible recklessness of miners. Small accidents, also, transpire at Ardeer in the repair of pipes. A drop of nitroglycerin which has secreted itself in a crack or crevice in the metal is sometimes struck by a hard tool, and costs a plumber one or more fingers.

These facts concerning dynamite are well known, and they are very reassuring. As you enter the train to leave Ardeer, however, the old habit of doubt reasserts itself. A bit of white fluff on your coat sleeve is viewed with the greatest suspicion. The question arises, "Is it cotton or gun-cotton?" Nerving yourself to the ordeal, you deliberately pick it off. You then carefully throw it out of the window to wreak its fell purpose, if it has one, on the landscape. Then you settle back with a vague desire to look at a thermometer. You have acquired a respect, an admiration, for any and all thermometers, which will abide with you to the end of your days.

The "Flying-Machine"

By Professor S. P. Langley

With illustrations made directly from Professor Langley's machine and approved by him.

June 1897

I HAVE been asked to prepare an account of some experiments I have conducted with flying-machines, built chiefly of steel, driven by steam-engines, and which have actually flown for considerable distances. There is in preparation a description of this work for the professional reader; but in view of the great general interest in it, and of the numerous unauthorized statements about it, it has seemed well to write provisionally the informal and popular account which is now given. The work has occupied so much of my life that I have presented what I have to say at present in narrative form.

By "flying-machine" is here meant something much heavier than the air, and entirely different in principle from the balloon, which floats only on account of its lightness, as a ship in water. Nature has made her flying-machine in the bird, which is nearly a thousand times as heavy as the air its bulk displaces, and only those who have tried to rival it know how inimitable her work is, for the "way of a bird in the air" remains as wonderful to us as it was to Solomon, and the sight of the bird has constantly held this wonder before men's eyes and in some men's minds, and kept the flame of hope from utter extinction, in spite of long disappointment. I well remember how, as a child, when lying in a New England pasture, I watched a hawk soaring far up in the blue, and sailing for a long time without any motion of its wings, as though it needed no work to sustain it, but was kept up there by some miracle. But, however sustained, I saw it sweep, in a few seconds of its leisurely flight, over a distance that to me was encumbered with every sort of obstacle, which did not exist for it. The wall over which I had climbed when I left the road, the ravine I had crossed, the patch of undergrowth through which I had pushed

Professor Langley's aërodrome in flight: A view from above.

my way—all these were nothing to the bird, and while the road had only tak‑ en me in one direction, the bird's level highway led everywhere, and opened the way into every nook and corner of the landscape. How wonderfully easy, too, was its flight! There was not a flutter of its pinions as it swept over the field, in a motion which seemed as effortless as that of its shadow.

After many years and in mature life, I was brought to think of these things again, and to ask myself whether the problem of artificial flight was as hopeless and as absurd as it was then thought to be. Nature had solved it, and why not man? Perhaps it was because he had begun at the wrong end, and attempted to construct machines to fly before knowing the principles on which flight rested. I turned for these principles to my books, and got no help. Sir Isaac Newton had indicated a rule for finding the resistance to advance through the air, which seemed, if correct, to call for enormous mechanical power, and a distinguished French mathematician had given a formula show‑ ing how rapidly the power must increase with the velocity of flight, and ac‑ cording to which a swallow, to attain a speed it is now known to reach, must be possessed of the strength of a man.

Remembering the effortless flight of the soaring bird, it seemed that the first thing to do was to discard rules which led to such results, and to com‑ mence new experiments, not to build a flying‑machine at once, but to find the principles upon which one should be built; to find, for instance, with certainty by direct trial how much horse-power was needed to sustain a surface of given weight by means of its motion through the air.

Having decided to look for myself at these questions, and at first hand, the apparatus for this preliminary investigation was installed at Allegheny, Pennsylvania, about ten years ago. It consisted of a "whirling table" of unprecedented size, mounted in the open air, and driven round by a steam-engine, so that the end of its revolving arm swept through a circumference of two hundred feet, at all speeds up to seventy miles an hour. At the end of this arm was placed the apparatus to be tested, and, among other things, this included surfaces disposed like wings, which were hung from the end of the arm and dragged through the air, till its resistance supported them as a kite is supported by the wind. One of the first things observed was that if it took a certain strain to sustain a properly disposed weight while it was stationary in the air, then not only to suspend it but to advance it rapidly at the same time, took less strain than in the first case. A plate of brass weighing one pound, for instance, was hung from the end of the arm by a spring, which was drawn out till it registered that pound weight when the arm was still. When the arm was in motion, with the spring pulling the plate after it, it might naturally be supposed that, as it was drawn faster, the pull would be greater, but the contrary was observed, for under these circumstances the spring *contracted*, till it registered less than an ounce. When the speed increased to that of a bird, the brass plate seemed to float on the air; and not only this, but taking into consideration both the strain and the velocity, it was found that absolutely less power was spent to make the plate move fast than slow, a result which seemed very extraordinary, since in all methods of land and water transport a high speed costs much

Professor S. P. Langley

From the painting by Robert Gordon Hardie, 1893.

more power than a slow one for the same distance.

These experiments were continued for three years, with the general conclusion that by simply moving any given weight of this form fast enough in a horizontal path it was possible to sustain it with less than one-twentieth of the power that Newton's rule called for. In particular it was proved that if we could insure horizontal flight without friction, about two hundred pounds of such plates could be moved through the air at the speed of an express train and sustained upon it, with the expenditure of one horse-power—sustained, that is, without any gas to lighten the weight, or by other means of flotation than the air over which it is made to run, as a swift skater runs safely over thin ice, or a skipping stone goes over water without sinking, till its speed is exhausted. This was saying that, so far as power alone was concerned, mechanical flight was theoretically possible with engines we could then build, since I was satisfied that boilers and engines could be constructed to weigh less than twenty pounds to the horse-power, and that one horse-power would, in theory at least, support nearly ten times that if the flight were *horizontal.* Almost everything, it will be noticed, depends on this, for if the flight is downward it will end at the ground, and if upward the machine will be climbing an invisible hill, with the same or a greater effort than every bicycler experiences with a real one. Speed, then, and this speed expended in a horizontal course, were the first two requisites. This was not saying that a flying-machine could be started from the ground, guided into such flight in any direction, and brought back to earth in safety. There was, then, something more than power needed—that is, skill to use it, and the reader should notice the distinction. Hitherto it had always been supposed that it was wholly the lack of mechanical power to fly which made mechanical flight impossible. The first stage of the investigation had shown how much, or rather how little, power was needed in theory for the horizontal flight of a given weight, and the second stage, which was now to be entered upon, was to show first how to procure this power with as little weight as possible, and, having it, how by its means to acquire this horizontal flight in practice—that is, how to acquire the *art* of flight or how to

Preparing to launch the aërodrome.

From a photograph by A. Graham Bell. Esq.

build a ship that could actually navigate the air.

One thing which was made clear by these preliminary experiments, and made clear nearly for the first time, was that if a surface be made to advance rapidly, we secure an essential advantage in our ability to support it. Clearly we want the advance to get from place to place; but it proves also to be the only practicable way of supporting the thing at all, to thus take advantage of the inertia of the air, and this point is so all-important that we will renew an old illustration of it. The idea in a vague sense is as ancient as classical times. Pope says:

"Swift Camilla scours the plain,

Flies o'er the unbending corn, and skims along the

main."

Now, is this really so in the sense that a Camilla, by running fast enough, could run over the tops of the corn? *If* she ran fast enough, yes; but the idea may be shown better by the analogous case of a skater who can glide safely over the thinnest ice if the speed is sufficient.

Think of a cake of ice of any small size, suppose a foot square. It possesses (like everything else in nature) inertia or resistance to displacement, and this will be less or more according to the mass moved. If the skater stands during a single second upon this small mass it will sink under him until he is perhaps waist-deep in the water, while a cake of the same width but twice the length will yield only about half as readily to his weight. On this he will sink only to his knees, we may suppose, while if we think of another cake ten times as long as the first—that is, one foot wide and ten feet long—we see that on this, during the same second, he will not sink above his feet. This is all plain enough; but now suppose the long cake to be divided into ten distinct portions, then it ought to be equally clear that the skater who glides over the whole in a second, distributes his weight over just as much ice as though all ten were in one solid piece. So it is with the air. Even the viewless air possesses inertia; it cannot be pushed aside without some effort; and while the portion which is directly under the airship would not keep it from falling several yards in the first second, if the ship goes forward so that it runs or treads on thousands of such portions in that time, it will sink in proportionately less degree; sink, perhaps, only through a fraction of an inch.

Speed, then, is indispensable here. A balloon, like a ship, will float over one spot in safety, but our flying-machine must be in motion to sustain itself, and in motion, in fact, before it can even begin to fly.

Perhaps we may more fully understand what is meant by looking at a boy's kite. Everyone knows that it is held by a string against the wind which sustains it, and that it falls in a calm. Most of us remember that even in a calm, if we run and draw it along, it will still keep up, for what is required is motion relative to the air, however obtained.

It can be obtained without the cord if the same pull is given by an engine and propellers strong enough to draw it, and light enough to be attached to and sustained by it. The stronger the pull and the quicker the motion, the heavier the kite may be made. It may be, instead of a sheet of paper, a sheet of metal even, like the plate of brass which

The aërodrome in flight, May 6, 1896. Two views from instantaneous photographs taken by A. Graham Bell, Esq.

has already been mentioned as seeming, when in rapid motion, to float upon the air, and, if it will make the principle involved more clear, the reader may think of our aërodrome as a great steel kite made to run fast enough over the air to sustain itself, whether in a calm or in a wind, by means of its propelling machinery, which takes the place of the string.

And now having the theory of the flight before us, let us come to the practice. The first thing will be to provide an engine of unprecedented lightness, that is to furnish the power. A few years ago an engine that developed a horse-power, weighed nearly as much as the actual horse did. We have got to begin by trying to make an engine which shall weigh, everything complete, boiler and all, not more than twenty pounds to the horse-power, and preferably less than ten; but even if we have done this very hard thing, we may be said to

The aërodrome in flight: A view from below.

have only fought our way up to an enormous difficulty, for the next question will be how to use the power it gives so as to get a horizontal flight. We must then consider through what means the power is to be applied when we get it, and whether we, shall, for instance, have wings or screws. At first it seems as though Nature must know best, and that since her flying models, birds, are exclusively employing wings, this is the thing for us; but perhaps this is not the case. If we had imitated the horse or the ox, and made the machine which draws our trains walk on legs, we should undoubtedly never have done as well as with the locomotive rolling on wheels; or if we had imitated the whale with its fins, we should not have had so good a boat as we now have in the steam-ship with the paddle-wheels or the screw, both of which are constructions that Nature never employs. This is so important a point that we will look at the way Nature got her models. Here is a human skeleton, and here one of a bird, drawn to the same scale. Apparently Nature made one out of the other, or both out of some common type, and the closer we look, the more curious the likeness appears.

Here is a wing from a soaring bird, here the same wing stripped of its feathers, and here the bones of a human arm, on the same scale. Now, on comparing them we see still more clearly than in the skeleton, that the bird's wing has developed out of something like our own arm. First comes the humerus, or principal bone of the upper arm, which is in the wing also. Next we see that the forearm of the bird re-peats the radius and ulna, or two bones of our own forearm, while our wrist and finger-bones are modified in the bird to carry the feathers, but are still there. To make the bird, then, Nature appears to have taken what material she had in stock, so to speak, and developed it into something that would do. It was all that Nature had to work on, and she has done wonderfully well with such unpromising material; but any one can see that our arms would not be the best thing to make flying-machines out of, and that there is no need of our starting there when we can start with something better and develop that. Flapping wings might be made on other principles, and perhaps will be found in future flying-machines, but the most promising thing to try seemed to me to be the screw propeller.

Some twenty years ago, Penaud, a Frenchman, made a toy, consisting of a flat, immovable sustaining wing surface, a flat tail, and a small propelling screw. He made the wing and tail out of paper or silk, and the propeller out of cork and feathers, and it was driv-en directly by strands of india-rubber twisted lamplighter fashion, and which turned the wheel as they un-twisted.

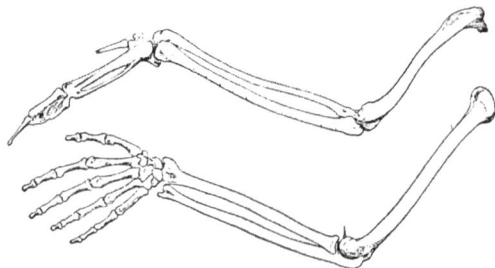

The great difficulty of the task of creating a flying-machine may be partly understood when it is stated that no machine in the whole history of invention, unless it were this toy of

The bones of a bird's wing and the bones of a human arm, drawn to the same scale, showing the close resemblance between them.

Penaud's, had ever, so far as I can learn, flown for even ten seconds; but some-thing that will actually fly must be had to teach the art of "balancing."

When experiments are made with models moving on a whirling table or running on a railroad track, these are *forced* to move horizontally and at the same time are held so that they cannot turn over; but in free flight there will be nothing to secure this, unless the airship is so adjusted in all its parts that it tends to move steadily and horizontally, and the acquisition of this adjust-ment or art of "balancing" in the air is an enormously difficult thing, and which, it will be seen later, took years to acquire.

My first experiments in it, then, were with models like these, but from them I got only a rude idea how to balance the future aërodrome, partly on account of the brevity of their flight, which only lasted a few seconds, partly on account of its irregularity. Although, then, much time and labor were spent by me on these, it was not possible to learn much about the balancing from them.

Thus it appeared that something which could give longer and steadier flights than india-rubber must be used as a motor, even for the preliminary trials, and calculations and experiments were made upon the use of compressed air, carbonic acid gas, electricity in primary and storage batteries, and numerous other contrivances, but all in vain. The gas-engine promised to be best ultimately, but nothing save steam gave any promise of immediate success in supporting a machine which would teach these conditions of flight by actual trial, for all were too heavy, weight being the great enemy. It was true also that the steam-driven model could not be properly constructed until the principal conditions of flight were learned, nor these be learned till the working model was experimented with, so that it seemed that the inventor was shut up in a sort of vicious circle.

However, it was necessary to begin in some way, or give up at the outset, and the construction began with a machine to be driven by a steam-engine, through the means of propeller wheels, somewhat like the twin screws of a modern steamship, but placed amidships, not at the stern. There were to be rigid and motionless wings, slightly inclined, like the surface of a kite, and a construction

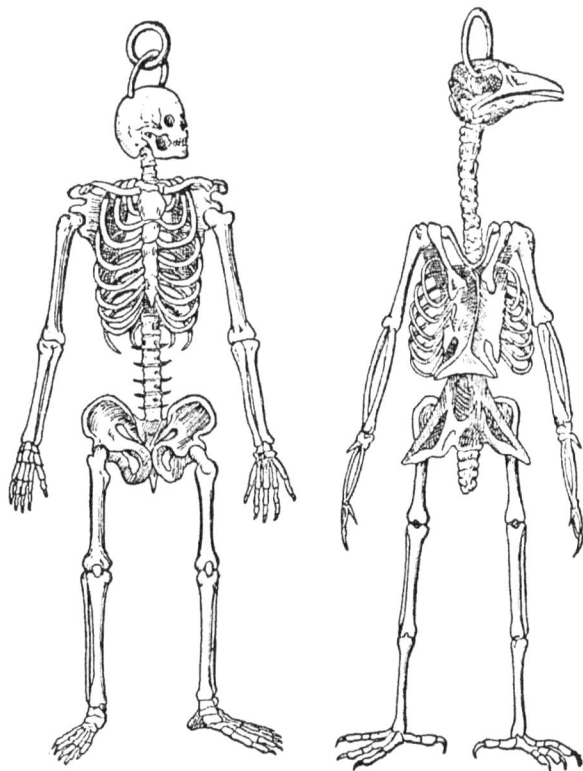

The skeleton of a man and the skeleton of a bird, drawn to the same scale, showing the curious likeness between them.

was made on this plan which gave, if much disappointment, a good deal of useful experience. It was intended to make a machine that would weigh twenty or twenty-five pounds, constructed of steel tubes. The engines were made with the best advice to be got (I am not an engineer); but while the boiler was a good deal too heavy, it was still too small to get up steam for the engines, which weighed about four pounds, and could have developed a horse-power if there were steam enough. This machine, which was to be moved by two propelling screws, was labored on for many months, with the result that the weight was constantly increasing beyond the estimate until, before it was done, the whole weighed over forty pounds, and yet could only get steam for about a half horse-power, which, after deductions for loss in transmission, would give not more than half that gain in actual thrust. It was clear that whatever pains it had cost, it must be abandoned.

This aërodrome[1] could not then have flown; but having learned from it the formidable difficulty of making such a thing light enough, another was constructed, which was made in the other extreme, with two engines to be driven by compressed air, the whole weighing but five or six pounds. The power proved insufficient. Then came another, with engines to use carbonic-acid gas, which failed from a similar cause. Then followed a small one to be run by steam, which gave some promise of success, but when tried indoors it was found to lift only about one-sixth of its own weight. In each of these the construction of the whole was re-modeled to get the greatest strength and lightness combined, but though each was an improvement on its predecessor, it seemed to become more and more doubtful whether it could ever be made sufficiently light, and whether the desired end could be reached at all.

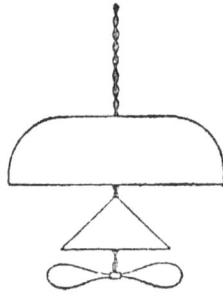

Penaud's flying toy

The chief obstacle proved to be not with the engines, which were made surprisingly light after sufficient experiment. The great difficulty was to make a boiler of almost no weight which would give steam enough, and this was a most wearying one. There must be also a certain amount of wing surface, and large wings weighed prohibitively; there must be a frame to hold all together, and the frame, if made strong enough, must yet weigh so little that it seemed impossible to make it. These were the difficulties that I still found myself in after two years of experiment, and it seemed at this stage again as if it must, after all, be given up as a hopeless task, for somehow the thing had to be built stronger and lighter yet. Now, in all ordinary construction, as in building a steamboat or a house, engineers have what they call a factor of safety. An iron column, for instance, will be made strong enough to hold five or ten times the weight that is ever going to be put upon it, but if we try anything of the kind here the construction will be too heavy to fly. Everything in the work has got to be so light as to be on the edge of breaking down and disaster, and when the breakdown comes all we can do is to find what is the weakest part and make that part stronger; and in this way work went on, week by week and month by month, constantly altering the form of construction so as to strengthen the weakest parts, until, to abridge a story which extended over years, it was finally brought nearly to the shape it is now, where the completed mechanism, furnishing over a horse-power, weighs collectively something less than seven pounds. This does not include water, the amount of which depends on how long we are to run; but the whole thing, as now constructed, boiler, fire-grate, and all that is required to turn out an actual horse-power and more, weighs something less than one one-hundredth part of what the horse himself does. I am here anticipating; but after these first three years something not greatly inferior to this was already reached, and so long ago as that, there had accordingly been secured mechanical power to fly, if that were all—but it is not all.

After that came years more of delay arising from other causes, and I can hardly repeat the long story of subsequent disappointment, which commenced with the first attempts at actual flight.

Mechanical power to fly was, as I say, obtained three years ago; the machine could lift itself if it ran along a railroad track, and it might seem as though, when it could lift itself, the problem was solved. I knew that it was far from solved, but felt that the point was reached where an attempt at actual free flight should be made, though the anticipated difficulties of this were of quite another order to those experienced in shop construction. It is enough to look up at the gulls or buzzards, soaring overhead, and to watch the incessant rocking and balancing which accompanies their gliding motion to apprehend that they find something more than mere strength of wing necessary, and that the machine would have need of something more than mechanical power, though what this something was, was not clear. It looked as though it might need a power like instinctive adaptation to the varying needs of each moment, something that even an intelligent steersman on board could hardly supply, but to find what this was a trial had to be made. The first difficulty seemed to be to make the initial flight in such conditions that the machine would not wreck itself at the outset, in its descent, and the first question was where to attempt to make the flight.

It became clear without much thought, that since the machine was at first unprovided with any means to save it from breakage on striking against the ground, it would be well, in the initial stage of the experiment, not to have it light on the ground at all, but on the water. As it was probable that, while skill in launching was being gained, and until after practice had made perfect, failures would occur, and as it was not desired to make any public exhibition of these, a great many places were examined along the shores of the Potomac, and on its high bluffs, which were condemned partly for their publicity, but partly for another reason. In the course of my experiments I had found out, among the infinite things pertaining to this problem, that the machine must begin to fly in the face of the wind, and just in the opposite way to a ship, which begins its voyage with the wind behind it. If the reader has ever noticed a soaring bird get upon the wing, he will see that it does so with the breeze against it, and thus whenever the aërodrome is cast into the air, it must face a wind which may happen to blow from the north, south, east, or west, and we had better not make the launching station a place like the bank of a river, where it can go only one way. It was necessary, then, to send it from something which could be turned in any direction, and taking this need in connection with the desirability that at first the airship should light in the water, there came at last the idea (which seems obvious enough when it is stated) of getting some kind of a barge or boat, and building a small structure upon it, which could house the aërodrome when not in use, and from whose flat roof it could be launched in any direction. Means for this were limited, but a little "scow" was procured, and on it was built a primitive sort of a house, one story high, and on the house a platform about ten feet higher, so that the top of the platform was about twenty feet from the water, and this was to be the place of the launch. This boat it was found necessary to take down the. river as much as thirty miles from Washington, where I then was,—since no suitable place could be found nearer,—to an island having a stretch of quiet water between it and the main shore; and here the first experiments in attempted flight developed difficulties of a new kind, difficulties which were partly anticipated, but which nobody would probably have conjectured would be of their actually formidable character, which was such as for a long time to prevent any trial being made at all. They arose partly out of the fact that even such a flying-machine as a soaring bird has to get up an artificial speed before it is on the wing. Some soaring birds do this by an initial run upon the ground, and even under the most urgent pressure cannot fly without it.

Take the following graphic description of the commencement of an eagle's flight (the writer was in Egypt, and the "sandy soil" was that of the banks of the Nile):

"An approach to within eighty yards aroused the king of birds from his apathy. He partly opened his enormous wings, but stirs not yet from his station. On gaining a few feet more he begins to *walk* away, with half-expanded but motionless wings. Now for the chance, fire! A charge of number three from eleven bore rattles audibly but ineffectively upon his densely feathered body; his walk increases to a run, he gathers speed with his slowly waving wings, and eventually leaves the ground. Rising at a gradual inclination, he mounts aloft and sails majestically away to his place of refuge in the Libyan range, distant at least five miles from where he rose. Some fragments of feathers denoted the spot where the shot had struck him. The marks of his claws were traceable in the sandy soil, as, at first with firm and decided digs, he forced his way, but as he lightened his body and increased his speed with the aid of his wings, the imprints of his talons gradually merged into long scratches. The measured distance from the point where these vanished, to the place where he had stood, proved that with all the stimulus that the shot must have given to his exertions, he had been compelled to run full twenty yards before he could raise himself from the earth."

We have not all had a chance to see this striking illustration of the necessity of getting up a preliminary speed before soaring, but many of us have disturbed wild ducks on the water and noticed them run along it, flapping their wings for some distance to get velocity before they can fly, and the necessity of the initial velocity is at least as great with our flying-machine as it is with a bird.

To get up this preliminary speed, many plans were proposed, one of which was to put the aërodrome on the deck of a steamboat and go faster and faster until the head wind lifted it off the deck. This sounds reasonable, but is absolutely impracticable, for when the aërodrome is set up anywhere in the open air we find that the very slightest wind will turn it over, unless it is firmly held. The whole must be in motion, but in motion from something to which it is held till that critical instant when it is set free as it springs into the air.

The house-boat was fitted with an apparatus for launching the aërodrome with a certain initial velocity, and was (in 1893) taken down the river and moored in the stretch of quiet water I have mentioned, the general features of the place being indicated on the accompanying map; and it was here that the first trials at launching were made, under the difficulties to which I have alluded.

Perhaps the reader will take patience to hear an abstract of a part of the diary of these trials, which commenced with a small aërodrome which had finally been built to weigh only about ten pounds, which had an engine of not quite one-half horsepower, and which could lift much more than was theoretically necessary to enable it to fly. The exact construction of this early aërodrome is unimportant, as it was replaced later by an improved one, of which a drawing is given, but it was the first outcome of the series of experiments which had occupied three years, though the disposition of its supporting surfaces, which should cause it to be properly balanced in the air and neither fly up nor down, had yet to be ascertained by trial.

What must still precede this trial was the provision of the apparatus for launching it into the air. It is a difficult thing to launch a ship, although gravity keeps it down upon the ways, but the problem here is that of launching a kind of ship which is as ready to go up into the air like a balloon as to go off sideways, and readier to do either than to go straight forward, as it is wanted to do, for though there is no gas in the flying-machine, its great extent of wing surface renders it something like an albatross on a ship's deck—the most unmanageable and helpless of creatures until it is in its proper element.

If there were an absolute calm, which never really happens, it would still be impracticable to launch it as a ship is launched, because the wind made by running it along would get under the wings and turn it over. But there is always more or less wind, and even the gentlest breeze was afterward found to make the airship unmanageable unless it was absolutely clamped down to whatever served to launch it, and when it was thus firmly clamped, as it must be at several distinct points, it was necessary that it should be released simultaneously at all these at the one critical instant that it was leaping into the air. This is another difficult condition, but that it is an indispensable one may be inferred from what has been said. In the first form of launching-piece this initial velocity was sought to be attained by a spring, which threw forward the supporting frame on which the aërodrome rested; but at this time the extreme susceptibility of the whole construction to injury from the wind, and the need of protecting it from even the gentlest breeze, had not been appreciated by experience. On November 18, 1893, the aërodrome had been taken down the river, and the whole day was spent in waiting for a calm, as the machine could not be held in position for launching for two seconds in the lightest breeze. The party returned to Washington and came down again on the 20th, and although it seemed that there was scarcely any movement in the air, what little remained was enough to make it impossible to maintain the aërodrome in position. It was let go, notwithstanding, and a portion struck against the edge of the launching-piece, and all fell into the water before it had an opportunity to fly.

On the 24th, another trip was made, and another day spent ineffectively on account of the wind. On the 27th there was a similar experience, and here four days and four (round-trip) journeys of sixty miles each had been spent without a single result. This may seem to be a trial of patience, but it was repeated in December, when five fruitless trips were made, and thus nine such trips were made in these two months, and but once was the aërodrome even attempted to be launched, and this attempt was attended with disaster. The principal cause lay, as I have said, in the unrecognized amount of difficulty introduced even by the very smallest wind, as a breeze of three or four miles an hour, hardly perceptible to the face, was enough to keep the airship from resting in place for the critical seconds preceding the launching.

If we remember that this is all irrespective of the fitness of the launching-piece itself, which at first did not get even a chance for trial, some of the difficulties may be better understood, and there were many others.

During most of the year of 1894 there was the same record of defeat. Five more trial trips were made in the spring and summer, during which various forms of launching apparatus were tried with varied forms of disaster. Then it was sought to hold the aërodrome out over the water and let it drop from the greatest attainable height, with the hope that it might acquire the requisite speed of advance before the water was reached. It will hardly be anticipated that it was found impracticable at first to simply let it drop, without something going wrong, but so it was, and it soon became evident that even were this not the case, a far greater time of fall was requisite for this method than that at command. The result was that in all these eleven months the aërodrome had not been launched, owing to difficulties which seem so slight that one who has not experienced them may wonder at the trouble they caused.

Finally, in October, 1894, an entirely new launching apparatus was completed, which embodied the dozen or more requisites, the need for which had been independently proved in this long process of trial and error. Among these was the primary one that it was capable of sending the aërodrome off at the requisite initial speed, in the face of a wind from whichever quarter it blew, and it had many more facilities which practice had proved indispensable.

This new launching-piece did its work in this respect effectively, and subsequent disaster was, at any rate, not due to it. But now a new series of failures took place, which could not be attributed to any defect of the launching apparatus, but to a cause which was at first obscure, for sometimes the aërodrome, when successfully launched, would dash down forward and into the water, and sometimes (under apparently identically like conditions) would sweep almost vertically upward in the air and fall back, thus behaving in entirely opposite ways, although the circumstances of flight seemed to be the same. The cause of this class of failure was finally found in the fact that as soon as the whole was upborne by the air, the wings yielded under the pressure which supported them, and were momentarily distorted from the form designed and which they appeared to possess. "Momentarily," but enough to cause the wind to catch the top, directing the flight downward, or under them, directing it upward, and to wreck the experiment. When the cause of the difficulty was found, the cure was not easy, for it was necessary to make these great sustaining surfaces rigid so that they could not bend, and to do this without making them heavy, since weight was still the enemy; and nearly a year passed in these experiments.

Has the reader enough of this tale of disaster? If so, he may be spared the account of what went on in the same way. Launch after launch was successively made. The wings were finally, and after infinite patience and labor, made at once light enough and strong enough to do the work, and now in the long struggle the way had been fought up to the face of the final difficulty, in which nearly a year more passed, for the all-important difficulty of balancing the aërodrome was now reached, where it could be discriminated from other preliminary ones, which have been alluded to, and which at first obscured it. If the reader will look at the hawk or any soaring bird, he will see that as it sails through the air without flapping the wing, there are hardly two consecutive seconds of its flight in which it is not swaying a little from side to side, lifting one wing or the other, or turning in a way that suggests an acrobat on a tight-rope, only that the bird uses its widely outstretched wings in place of the pole.

There is something, then, which is difficult even for the bird, in this act of balancing. In fact, he is sailing so close to the wind in order to fly at all, that if he dips his head but the least he will catch the wind on the top of his wing and fall, as I have seen gulls do, when they have literally tumbled toward the water before they could recover themselves.

Beside this, there must be some provision for guarding against the incessant, irregular currents of the wind, for the wind as a whole—and this is a point of prime importance—is not a thing moving along all-of-a-piece, like water in the Gulf Stream. Far from it. The wind, when we come to study it, as we have to do here, is found to be made of innumerable currents and countercurrents which exist altogether and simultaneously in the gentlest breeze, which is in reality going fifty ways at once, although, as a whole, it may come from the east or the west; and if we could see it, it would be something like seeing the rapids below Niagara, where there is an infinite variety of motion in the parts, although there is a common movement of the stream as a whole.

All this has to be provided for in our mechanical bird, which has neither intelligence nor instinct, without which, although there be all the power of the engines requisite, all the rigidity of wing, all the requisite initial velocity, it still cannot fly. This is what is meant by balancing, or the disposal of the parts, so that the airship will have a position of equilibrium into which it tends to fall when it is disturbed, and which will enable it to move of its own volition, as it were, in a horizontal course.

Now the reader may be prepared to look at the apparatus which finally has flown. (See diagram on page 80.) In the completed form we see two pairs of wings, each slightly curved, each attached to a long steel rod which supports them both, and from which depends the body of the machine, in which are the boilers, the engines, the machinery, and the propeller wheels, these latter being not in the position of those of an ocean steamer, but more nearly amidships. They are made sometimes of wood, sometimes of steel and canvas, and are between three and four feet in diameter.

The hull itself is formed of steel tubing; the front portion is closed by a sheathing of metal which hides from view the fire-grate and apparatus for heating, but allows us to see a little of the coils of the boiler and all of the relatively large smoke-stack in which it ends. The conical vessel in front is an empty float, whose use is to keep the whole from sinking if it should fall in the water.

This boiler supplies steam for an engine of between one and one and one-half horsepower, and, with its fire-grate, weighs a little over five pounds. This weight is exclusive of that of the engine, which weighs, with all its moving

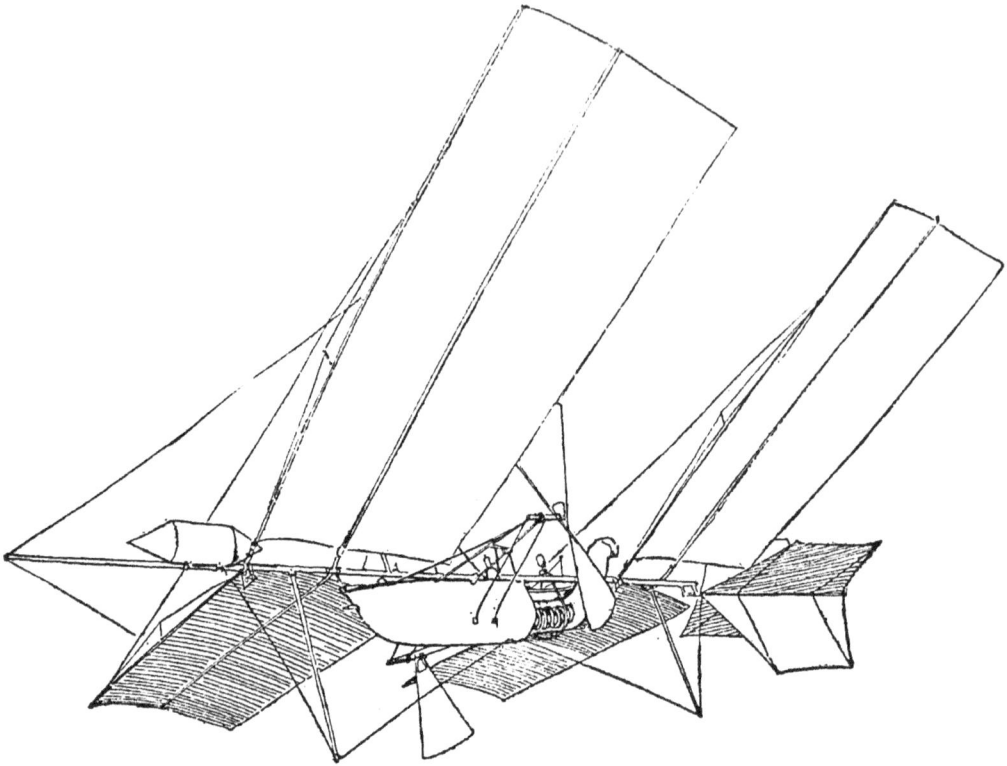

Diagram of the aërodrome.

parts, but twenty-six ounces. Its duty is to drive the propeller wheels, which it does at rates varying from 800 to 1,200, or even more, turns a minute, the highest number being reached when the whole is speeding freely ahead.

The rudder, it will be noticed, is of a shape very unlike that of a ship, for it is adapted both for vertical and horizontal steering. It is impossible within the limits of such an article as this, however, to give an intelligible account of the manner in which it performs its automatic function. Sufficient it is to say that it does perform it.

The width of the wings from tip to tip is between twelve and thirteen feet, and the length of the whole about sixteen feet. The weight is nearly thirty pounds, of which about one-fourth is contained in the machinery. The engine and boilers are constructed with an almost single eye to economy of weight, not of force, and are very wasteful of steam, of which they spend their own weight in five minutes. This steam might all be recondensed and the water re-used by proper condensing apparatus, but this cannot be easily introduced in so small a scale of construction. With it the time of flight might be hours instead of minutes, but without it the flight (of the present aërodrome) is limited to about five minutes, though in that time, as will be seen presently, it can go some miles; but owing to the danger of its leaving the surface of the water for that of the land, and wrecking itself on shore, the time of flight is limited designedly to less than two minutes.

I have spared the reader an account of numberless delays, from continuous accidents and from failures in attempted flights, which prevented a single entirely satisfactory one during nearly three years after a machine with power to fly had been attained. It is true that the aërodrome maintained itself in the air at many times, but some disaster had so often intervened to prevent a complete flight that the most persistent hope must at some time have yielded. On the 6th of May of last year I had journeyed, perhaps for the twentieth time, to the distant river station, and recommenced the weary routine of another launch, with very moderate expectation indeed; and when, on that, to me, memorable afternoon the signal was given and the aërodrome sprang into the air, [2] I watched it from the shore with hardly a hope that the long series of accidents had come to a close. And yet it had, and for the first time the aërodrome swept continuously through the air like a living thing, and as second after second passed on the face of the stop-watch, until a minute had gone by, and it still flew on, and as I heard the cheering of the few spectators, I felt that something had been accomplished at last, for never in any part of the world, or in any period, had any machine of man's construction sustained itself in the air before for even half of this brief time. Still the aërodrome went on in a rising course until, at the end of a minute and a half (for which time only it was provided with fuel and water), it had accomplished a little over half a mile, and now it settled rather than fell into the river with a gentle descent. It was immediately taken out and flown again with equal success, nor was there anything to indicate that it might not have flown indefinitely except for the limit put upon it.

I was accompanied by my friend, Mr. Alexander Graham Bell, who not only witnessed the flight, but took the instantaneous photograph of it which has been given. He spoke of it in a communication to the Institute of France in the following terms:

> Through the courtesy of Mr. S. P. Langley, Secretary of the Smithsonian Institution, I have had on various occasions the privilege of witnessing his experiments with aërodromes, and especially the remarkable success attained by him in experiments made on the Potomac River on Wednesday, May 6, which led me to urge him to make public some of these results.

I had the pleasure of witnessing the successful flight of some of these aërodromes more than a year ago, but Professor Langley's reluctance to make the results public at that time prevented me from asking him, as I have done since, to let me give an account of what I saw.

On the date named, two ascensions were made by the aëro-drome, or so-called "flying-machine," which I will not describe here further than to say that it appeared to me to be built al-most entirely of metal, and driven by a steam-engine which I have understood was carrying fuel and a water-supply for a brief period, and which was of extraordinary lightness.

The absolute weight of the aërodrome, including that of the en-gine and all appurtenances, was, as I was told, about twenty-five pounds, and the distance, from tip to tip, of the supporting surfaces was, as I observed, about twelve or fourteen feet.

The method of propulsion was by aerial screw propellers, and there was no gas or other aid for lifting it in the air except its own internal energy.

On the occasion referred to, the aërodrome, at a given signal, started from a platform about twenty feet above the water, and rose at first directly in the face of the wind, moving at all times with remarkable steadiness, and subsequently swinging around in large curves of, perhaps, a hundred yards in diameter, and continually ascending until its steam was exhausted, when, at a lapse of about a minute and a half, and at a height which I judged to be between eighty and one hundred feet in the air, the wheels ceased turning, and the machine, deprived of the aid of its propellers, to my surprise did not fall, but settled down so softly and gently that it touched the water without the least shock, and was in fact immediately ready for another trial.

In the second trial, which followed directly, it repeated in nearly every respect the actions of the first, except that the direction of its course was different. It ascended again in the face of the wind, afterwards moving steadily and continually in large curves, accompanied with a rising motion and a lateral advance. Its motion was, in fact, so steady that I think a glass of water on its surface would have remained unspilled. When the steam gave out again, it repeated for a second time the experience of the first trial when the steam had ceased, and settled gently and easily down. What height it reached at this trial I cannot say, as I was not so favorably placed as in the first; but I had occasion to notice that this time its course took it over a wooded promontory, and I was relieved of some apprehension in seeing that it was already so high as to pass the tree-tops by twenty or thirty feet. It reached the water one minute and thirty-one seconds from the time it started, at a measured distance of over nine hundred feet from the point at which it rose.

This, however, was by no means the length of its flight. I estimated from the diameter of the curve described, from the number of turns of the propellers as given by the automatic counter, after due allowance for slip, and from other measures, that the actual length of flight on each occasion was slightly over three thousand feet. It is at least safe to say that each exceeded half an English mile.

From the time and distance it will be noticed that the velocity was between twenty and twenty-five miles an hour, in a course which was constantly taking it "up hill." I may add that on a previous occasion I have seen a far higher velocity attained by the same aërodrome when its course was horizontal.

I have no desire to enter into detail further than I have done, but I cannot but add that it seems to me that no one who was present on this interesting occasion could have failed to recognize that the practicability of mechanical flight had been demonstrated.

Alexander Graham Bell.

On November 28th I witnessed, with another aërodrome of somewhat similar construction, a rather longer flight, in which it traversed about three-quarters of a mile, and descended with equal safety. In this the speed was greater, or about thirty miles an hour. We may live to see airships a common sight, but habit has not dulled the edge of wonder, and I wish that the reader could have witnessed the actual spectacle. "It looked like a miracle," said one who saw it, and the photograph, though taken from the original, conveys but imperfectly the impression given by the flight itself.

And now, it may be asked, what has been done? This has been done: a "flying-machine," so long a type for ridicule, has really flown; it has demonstrated its practicability in the only satisfactory way—by actually flying, and by doing this again and again, under conditions which leave no doubt.

There is no room here to enter on the consideration of the construction of larger machines, or to offer the reasons for believing that they may be built to remain for days in the air, or to travel at speeds higher than any with which we are familiar; neither is there room to enter on a consideration of their commercial value, or of those applications which will probably first come in the arts of war rather than those of peace; but we may at least see that these may be such as to change the whole conditions of warfare, when each of two opposing hosts will have its every movement known to the other, when no lines of fortification will keep out the foe, and when the difficulties of defending a country against an attacking enemy in the air will be such that we may hope that this will hasten rather than retard the coming of the day when war shall cease.

I have thus far had only a purely scientific interest in the results of these labors. Perhaps if it could have been foreseen at the outset how much labor there was to be, how much of life would be given to it, and how much care, I might have hesitated to enter upon it at all. And now reward must be looked for, if reward there be, in the knowledge that I have done the best I could in a difficult task, with results which it may be hoped will be useful to others. I have brought to a close the portion of the work which seemed to be specially mine—the demonstration of the practicability of mechanical flight—and for the next stage, which is the commercial and practical development of the idea, it is probable that the world may look to others. The world, indeed, will be supine if it do not realize that a new possibility has come to it, and that the great universal highway overhead is now soon to be opened.

[1] Aërodrome, from words signifying air-runner, the running over the air being the essence of its plan.

[2] This illustration, from an instantaneous photograph by Mr. Bell, shows the machine after Sir Reed, who was in charge of the launch (and to whom a great deal of the construction of the aërodrome is due), has released it, and when it is in the first instant of its aerial journey.

The Automobile in Common Use

What It Costs.—How It Is Operated. —What It Will Do.

By Ray Stannard Baker

July 1899

YESTERDAY, a mere mechanical wonder fresh from the hand of the inventor; today, a gigantic industry on two continents—that is the history in brief of the motor vehicle. Five years ago there were not thirty self-propelled carriages in practical use in all the world. A year ago there were not thirty in America. And yet between the 1st of January and the 1st of May, 1899, companies with the enormous aggregate capitalization of more than $388,000,000 have been organized in New York, Boston, Chicago, and Philadelphia for the sole purpose of manufacturing and operating these new vehicles. At least eighty establishments are now actually engaged in building carriages, coaches, tricycles, delivery wagons, and trucks, representing no fewer than 200 different types of vehicles, with nearly half as many methods of propulsion. Most of these concerns are far behind in their orders, and several of them are working day and night. A hundred electric cabs are plying familiarly on the streets of New York, and 200 more are being rushed to completion in order to supply the popular demand for horseless locomotion. At least two score of delivery wagons, propelled chiefly by electricity, are in operation in American cities, and the private conveyances of various makes will number well into the hundreds. A motor ambulance is in operation in Chicago; motor trucks are at work in several different cities; a motor gun-carriage for use in the army will be ready for service in the summer. The Santa Fe railroad has ordered a number of horseless coaches for an Arizona

A typical motor truck, motive power, compressed air.

mountain route, and at least two cities are using self-propelled fire-engines. A trip of 720 miles, from Cleveland to New York, over all kinds of country roads, has actually been made in a gasoline carriage, and an enthusiastic automobile traveler is now on his way from New England to San Francisco. And all of these doings are chronicled in a weekly journal devoted exclusively to the new industry.

These are a few of the important things which have been accomplished in America almost within the year. Never before has Yankee genius and enterprise created an important business interest in so short a time. The experimental plaything has become a practical necessary. And yet the motor vehicle in America is in its babyhood compared with what it is in France and England. Here it has hardly passed the stage of promotion and promise; there it has become an established and powerful factor in the common affairs of life, as well as a fashionable fad. France has an automobile club numbering 1,700 members. At its last exhibition 1,100 vehicles were shown, representing every con-

Models of the motor ambulance, motor tricycle, and motor omnibus now coming into use.

ceivable model, from milk-wagons to fashionable broughams and the huge brakes of De Dion and Bouton, which carry almost as many passengers as a railroad car. Some of the expert *chauffeurs* of Paris have ridden thousands of miles in their road wagons, have climbed mountains and raced through half of Europe, meeting new accidents, facing new adventures, and using strange new devices for which names have yet to be coined. In Paris, electric motor cabs are becoming quite as familiar as the old-fashioned horse cabs. Before the opening of the Paris Exposition, 1,200 of them will be in operation. In the country districts thousands of grocers, milkmen, market-men, and peddlers are the engineers of their own gasoline carts.

A French statistician has given some significant figures as to the enormous increase of the horse-slaughtering industry in Paris during the past two years, and he lays it largely to the thousands of motor vehicles which are making the horse more valuable for ragouts than for racing. The august French Academy has paused in its consideration of literature and art, to take cognizance of the motor vehicle, and has bestowed upon it the formal name of "automobile," which it expects the entire world to adopt. The French law has quietly absorbed its unfamiliar terms, and has decreed that every vehicle must be registered in its own commune, the same as a horse and carriage; it has laid down formal articles for the regulation of builders and operators, and provided for races and speed limits. The French Minister of War has numbered and described every vehicle in the republic, and has quietly arranged to seize them all for military purposes when France shall go to war. In this way the motor vehicle in France has assumed the settled importance of a governmental institution, as well as a great business industry.

England has not gone so far as France with the automobile, and yet it has several powerful associations devoted to its development, and a large number of vehicles in actual use. With his hard-headed, practical business sense, the Englishman is looking with greater care and interest into the development of the trucking vehicle, for carrying heavy loads, than to the lighter pleasure carriage. He has an eye to the enormous freight-rates of his railroads and to the crowded condition of his narrow streets. One successful exhibition of auto-trucks has already been held in Liverpool, under the auspices of the Self-Propelled Traffic Association, and a second, which is already anticipated with the keenest interest, will take place next August.

In general, France leads in gasoline vehicles, and England in steam vehicles, while America, as was to be expected, is far in the lead in electrical conveyances of all kinds. Belgium and Germany, and to some extent Austria, are also experimenting with more or less success, but no such progress has been made in these countries as in France. Spain rubbed its eyes last spring at the sight of its first motor vehicle, which rolled through Madrid with half a dozen little policemen careering after it. Indeed, the new industry is everywhere awakening the most extraordinary interest among all classes of people.

And yet the great public is far from feeling familiar with the motor vehicle. The prospective buyer, and there are many thousands of him in America, is at once confronted with the bewildering variety of models which the manufacturers place before him. He discovers that there are the most pronounced variations in price, cost of maintenance, speed, ease of management, and general efficiency.

It was with the idea of clearing up this confusion and giving some exact conception of what the motor vehicle of to-day really is, what it can do, what it costs, and what may be expected of it in the future, that I visited and talked with a number of the most prominent American manufacturers.

Fetching the doctor. Already physicians have found the automobile of special service to them.

In a general way, it may be said that the best modern motor vehicle, whatever its propelling power, is practically noiseless and odorless and nearly free from vibrations. It is still heavy and clumsy in appearance, although it is lighter than the present means of conveyance when the weight of the horse or horses is counted in with the carriage. And invention will soon lighten it still further. It cannot possibly explode. It will climb all ordinary hills, and on the level it will give all speeds from two miles an hour up to twenty or more. Its mechanism has been made so simple that any one can learn to manage it in an hour or two. And yet it is mechanism; and intelligence, coolness, and caution are required to manage a motor vehicle in a crowded street. The operator must combine the intelligence of the driver with that of the horse, and he does not appreciate the almost human sagacity of that despised animal until he has tried to steer a motor vehicle down Fifth Avenue on a sunny afternoon.

Six different motive powers are now actually employed in this country: electricity, gasoline, steam, compressed air, carbonic-acid gas, and alcohol. The first three of these have been practically applied with great success; all

A roadhouse scene, showing types of automobiles already in use.

the others are more or less in the experimental stage.

The electric vehicle, which has had its most successful development in this country, has its well-defined advantages and disadvantages. It is simpler in construction and more easily managed than any other vehicle: one manufacturer calls it "fool proof." It is wholly without odor or vibrations and practically noiseless. It can make any permissible rate of speed, and climb any hill up to a twenty per cent, grade. On the other hand, it is immensely heavy owing to the use of storage batteries; it can run only a limited distance without recharging, and it requires a moderately smooth road. In cost it is the most expensive of all vehicles. And yet for city use, where a constant supply of electricity can be had, electrical cabs, carriages, and delivery wagons have demonstrated their remarkable practicability.

The vital feature of the electric vehicle is the storage battery, which weighs from 500 to 1,500 pounds, the entire weight of the vehicles varying from about 900 to 4,000 pounds. A phaeton for ordinary use in carrying two people will weigh upwards of a ton, with a battery of 900 pounds. This immense weight requires exceedingly rigid construction and high-grade, expensive tires. The electrical current is easily controlled by means of a lever under the hand of the driver, the propelling machinery being comparatively simple. When the battery is nearly empty, it may be recharged at any electric-lighting station by the insertion of a plug, the time required varying from two to three hours. Or, if the owner prefers, he can own his own charging-plant and generate his own electricity; it will cost him from $500 to $700. The current not only operates the vehicle, but it lights the lamps, rings the gong, and in cabs and broughams actuates a push-button arrangement for communication between passenger and driver. The limit of travel without recharging is from twenty to thirty miles. Mr. C. E. Woods, a leading manufacturer, gives the cost of maintenance

Automobile tourists

of storage batteries per year as varying from $50 for light buggies to $300 for heavy omnibuses, the entire cost of operation being from three-quarters of a cent to four cents a mile. A good electric carriage for family use cannot be obtained for much less than $2,000, although one or two manufacturers advertise runabouts and buggies at from $750 to $1,500. An omnibus costs from $3,000 to $4,000. The Columbia Automobile Company has made an interesting comparison showing the cost of horse and electric delivery wagons:

FIRST COST.

HORSE WAGONS.		ELECTRIC WAGONS.	
Wagon	$380.00		
Two horses at $125	250.00	Electric wagon complete	$2,250.00
Harnesses	75.00		
	$705.00		

MAINTENANCE PER YEAR.

HORSE WAGONS.		ELECTRIC WAGONS.	
Interest on investment at 5 per cent	$35.25	Interest on investment at 5 per cent	$112.50
Stabling two horses at $38.50 both or $18.25 each, per month	438.00	Cost of electric current at ordinary central station, rates for 12,000 miles per year	300.00
Shoeing two horses	30.00		$412.50
Harness repairs, two horses	20.00		
	$523.25		

Or:

Interest	$112.50
Current, if generated in private plant	21.30
	$133.80

"In this table we omitted to mention repairs or the expense of a driver," the Columbia people said, "because we calculate that they are the same in both cases. And battery deterioration will offset horse deterioration. But in using the electric vehicle all stable odors and flies are done away with, and a second man is never necessary to 'watch the horse.' Moreover, an electric wagon can be kept in a quarter of the usual stable space, or even in the store itself."

The company which operates the electric cab system in New York has a most extensive charging-plant. Two batteries are provided for each vehicle, so that, when one is empty, it may be removed by the huge fingers of a traveling crane, placed on a long table, and recharged at leisure, while a completely filled battery is introduced in its place. This change takes only a few minutes, and the cab can be used continuously day and night.

The "lightning cabby" is a product of the new industry. He wears a blue uniform somewhat resembling that of a fireman, and he is a cool-headed, intelligent fellow, who can make ten miles an hour in a crowded street without once catching the suspicious eye of a policeman. Most of the "cabbies" have had previous experience as drivers, but they are given a very thorough training before they are allowed to venture on the streets with a vehicle of their own. A special instructor's cab is in use by the company. It has a flaring front platform with a solid wooden bumper, so that it may crash into a stone curb or run down a lamp-post without injury. The new man perches himself on the seat behind, and the instructor takes his place inside, where he is provided with a special arrangement for cutting off the current or applying the brakes, should the vehicle escape from the control of the learner. It usually takes a week to train a new man so that he can manage all the brakes and levers with perfect presence of mind. Both of his hands and both of his feet are fully employed. With his left hand he manages the power lever, pushing it forward one notch at a time to increase the speed. With his right hand he controls the steering-lever, which, by the way, turns the rear wheels and not the front ones, as is done with horse-propelled vehicles. His left heel is on the emergency switch, and his left toes ring the gong. With his right heel he turns the reversing-switch, and he can apply the brake with either his right or his left foot. When he wishes to turn on the lights, he presses, a button under the edge of the seat. Hence, he is very fully employed, both mentally and physically. He can't go to sleep and let the old horse carry him home.

In France the system of instruction for drivers or *chauffeurs* (stokers), as they are called, is much more complicated and extensive, but hardly more

The training course for automobile drivers at Aubervilliers, near Paris.

The course, besides being obstructed by the dummy figures shown in the picture, is strewn with paper bricks, and thus becomes as severe a test as possible of the skill of the motorman.

thorough. There the cab company has prepared a 700-yard course up hill and down, and paved it alternately with cobbles, asphalt, wooden blocks, and macadam, so as to give the incipient "cabby" experience in every difficulty which he will meet in the streets of Paris. Upon the inclines are placed numerous lay figures, made of iron—a typical Parisian nurse-maid with a bassinet; a bicycle rider; an old gentleman, presumably deaf, who is not spry in getting out of the way; a dog or two, and paper bricks galore. Down through this throng must the motorman thread his way and clang his gong, and he is not considered proficient until he can course the full length of the "Rue de Magdebourg," as the cabbies call it, without so much as overturning a single pastry cook's boy or crushing a dummy brick.

New York cabs will run twenty miles without recharging. But it is not at all infrequent for a new man to have his vehicle stop suddenly and most unexpectedly; the current deserts him before he knows it. He must let the central office know at once, and the ambulance cab comes spinning out, hooks to the helpless vehicle, and drags it in to the charging-station. The company expects soon to have ten charging-stations in operation in various parts of the city, so that a cab will never have far to go for a new charge of electricity. Indeed, all the manufacturers of electrical vehicles speak with confidence of the day when the whole of the United States will be as thoroughly sprinkled with electric charging-stations as it is to-day with bicycle road-houses. One manufacturer has already issued lists of hundreds of central stations throughout New England, New York, and other Eastern States where automobiles may be provided with power.

It is not hard to imagine what a country touring-station will be like on a sunny summer afternoon some five or ten years hence. Long rows of vehicles will stand backed up comfortably to the charging-bars, each with its electric plug filling the battery with power. The owners will be lolling at the tables on the verandas of the nearby road-house. Men with repair kits will bustle about, tightening up a nut here, oiling this bearing, and regulating that gear. From a long rubber tube compressed air will be hissing into pneumatic tires. There will also be many gasoline carts and road-wagons and tricycles, and they, too, will need repairs and pumping, and their owners will employ themselves busily in filling their little tin cans with gasoline, recharging their tanks, refilling the water-jackets, and looking to the working of their sparking devices. And then there will be boys selling peanuts, arnica, and court-plasters, and undoubtedly a cynical old farmer or two with a pair of ambling mares to carry home such of these new-fangled vehicles as may become hopelessly indisposed. Add to this bustling assembly of amateur "self-propellers" a host of bicycle riders—for there will doubtless be as many bicycles in those days as ever—and it will be a sight to awaken every serious-minded horse to an uneasy consideration of his future.

Nor is this dream so far from being a picture of actual conditions. In Belgium a company has recently been formed to establish electric posting stations. Its promoters plan to have a bar and restaurant connected with the charging-plant, a regular medical attendant, and an expert mechanic who will know how to remedy all the ills of motor vehicles. In the larger cities the time must soon come when there will be coin-in-the-slot "hydrants" for electricity at many public places from which owners of vehicles may charge their batteries while they wait.

The new electric cabs are unquestionably immensely popular as fashionable conveyances. A number of the wealthy people of New York, including Mr. Frank Gould, Mr. Cornelius Vanderbilt, Mr. O. H. P. Belmont, and Mr. Richard McCurdy, have a cab or brougham and driver constantly on call at the home station of the company, for which they pay at the rate of $180 a month. Several prominent physicians are similarly provided, motor vehicles being especially adapted to the varied necessities of a physician's practice. A motor vehicle is always ready at a moment's notice—it does not have to be harnessed. It can work twenty-four hours a day. When it is left in the street outside, the doctor takes with him a little brass plug, or key, without which the vehicle cannot run away or be moved or stolen. And, moreover, it is swifter by half than the ordinary means of locomotion, so that in emergency cases it may mean the saving of a life. One New York physician recently put an electric cab to a most extraordinary use. His patient had a broken arm, and he wished to photograph the fracture with Roentgen rays, but there was no source of electricity available in the residence of the patient. So he made a connection with the battery in his cab, which stood at the door, the rays were promptly applied, and the injury was located.

While the electric vehicle has been winning plaudits for its work in the cities, where pavements are smooth and hard, the gasoline vehicle has been equally successful both in the city and in the country. For ordinary use the gasoline-propelled vehicle has many important advantages. It is much lighter than the electric vehicle; it requires no charging-station, gasoline being obtainable at every cross-roads store; and it is moderately cheap. All of the famous long-distance races and rides in Europe have been made in gasoline vehicles. On the other hand, most of the gasoline vehicles are subject to slight vibrations due to the motor, and it is almost impossible to do away entirely with the unpleasant odors of burnt gases. Gasoline vehicles are never self-starting, it being necessary to give the piston an initial impulse by hand. In general, also, they are not as simple of management as the electric vehicle; there is more machinery to understand and to operate, and more care is necessary to keep it in order. But when once the details are mastered, the traveler can go almost anywhere on earth with his gasoline carriage, up hill and down, over the roughest roads, through mud and snow, a law unto himself. He can make almost any speed he chooses. It is said that Baron De Knyff, of Paris, made fifty miles an hour for a short run, and Count Chesset-Loubat has surpassed even this record.

The power principle of the gasoline vehicle is very simple. It is a well-known fact that, when gasoline is mixed with air in proper proportions and ignited, it explodes violently. By admitting this mixture at the end or head of the engine cylinder, and exploding it at the proper moment, the piston is driven violently forward, and then, by the action of the fly-wheel or an equivalent device, it is forced back again, and the motor is kept in motion. Most gasoline engines are of what is known as the four-cycle variety. During the first impulse of the piston the vapor.is drawn into the end of the cylinder, during the second it is compressed by the return of the piston, in the third it is exploded, and in the fourth the products of the combustion are driven out, and the end of the cylinder is ready for another charge. The explosion of the gas is produced in the most approved motors by means of an electric spark, there being no fire anywhere connected with the machine. Owing to the constant compression of the gases and the succeeding explosions, a gasoline motor becomes highly heated, and in order to maintain a normal temperature, it must be provided with a

A motor tally-ho, propelled by stored electricity.

jacket of cold water, or a peculiar ribbed arrangement of iron for increasing the radiating surface. A vast number of ingenious devices are used for making all of these processes as simple as possible. One motor is so arranged that no igniter is necessary, the gas being compressed in the cylinder to such a degree that it explodes of its own heat, thereby doing away entirely with electricity or any other sparking-device. In France most of the gasoline vehicles are still provided with what are known as "carburetters," or small chambers where the gas and air are mixed in the proper proportions and heated before they are driven into the cylinder. In this country carburetters have been largely done away with, the gas being mixed as it passes into the cylinder.

Every driver of a gasoline vehicle must know these general facts about the mechanism of his motor. He must know how to fill the gasoline and water tanks, how to replenish or regulate the battery which ignites the gas, and he must understand the ordinary processes of cleaning and oiling machinery. When he is ready to start, he must connect up the sparking-device and turn the wheel controlling the piston until the explosions begin. After that, he must see that the valves which admit the air and the gas are carefully adjusted, so that the mixture is admitted to the cylinder in the proper proportion, and then he is ready to go ahead, steering and controlling his engine by means of levers, and operating the brake and gong with his feet. All gasoline vehicles are provided with several appliances for stopping besides the ordinary brake, so that there is practically no possible danger of a runaway. The Duryea vehicle, for instance, has no fewer than five different means of turning off the power of the motor, all within convenient reach. The secretary of the company that manufactures this vehicle told me that he had often stopped his carryall within twenty feet when going at a speed of twenty miles an hour, without great inconvenience to the passengers. By a clever arrangement for changing gearings the gasoline vehicle can be made to ascend almost any hill, and it can be turned in half the space necessary for a horse vehicle.

It is astonishing how little fuel it takes to run a gasoline vehicle. Mr. Fischer, of the American Motor Company, showed me a phaeton, weighing 700 pounds, which, he said, would run 100 miles on five gallons of gasoline, a bare half-dollar's worth. A tricycle manufactured by the same company, weighing 150 pounds, will run eighty miles on three pints of gasoline.

Gasoline vehicles vary in cost over an even wider range than electrical vehicles. A tricycle can be obtained as low as $350, while an omnibus may cost well into the thousands. A first-class road carriage built with all the latest improvements and highly serviceable in every respect can be obtained for $1,000. At this price, the manufacturers assert that gasoline power is much cheaper than horse power. Mr. A. S. Winslow, of the National Motor Carriage Company, has made some interesting comparisons, based on an average daily run of twenty-five miles for five years—more than the maximum endurance of a first-class horse. His estimates represent ordinary city conditions, and rate the cost of the gasoline used at one-half cent a mile:

"At the end of five years," said Mr. Winslow, "the motor vehicle should be in reasonably good condition, while the value of the horse and carriage would be doubtful. There is always the possibility that at least one of the horses may die in five years, while the motor vehicle can always be repaired at a comparatively nominal cost. But even assuming that the relative value of each is the same at the end of five years, the cost of actual maintenance during that period would be $1,306.50 for the motor vehicle and $2,280 for the horse and vehicle, or $973.50 in favor of the motor vehicle. This comparison is really doing more than justice to the horse, because a motor vehicle can do the work of three horses without injury."

GASOLINE MOTOR VEHICLE.

Original cost of the vehicle	$1,000.00
Cost of operation, 1 cent per mile, twenty-five miles per day	456.50
New sets of tires, during five years	100.00
Repairs on motor and vehicle	150.00
Painting vehicle four times	100.00
Storing and care of vehicle, $100.00 per year	500.00
	$2,300.50

HORSE AND VEHICLE.

Original cost of horse, harness, and vehicle	$500.00
Cost of keeping horse, $30.00 per month, five years	1.800.00
Repairs on vehicle, including rubber tires	150.00
Shoeing horse, $3.00 per month, five years	180.00
Repairs on harness, $10.00 per year	50.00
Painting vehicle four times	100.00
	$2,780.00

Steam has been successfully applied to the heavier grades of vehicles, notably trucks, fire-engines, and omnibuses; and two or three American manufacturers have gone still further, and have produced light and natty steam buggies and runabouts, and even steam tricycles. Steam vehicles are easily started and stopped, and fuel and water are always readily obtainable; but there is also the disadvantage of a slight cloud of steam escaping from the exhaust, accompanied by more or less noise. Moreover, in many States there are regulations (mostly unenforced in the case of motor vehicles) against the operation of steam engines except by licensed engineers, and it is probable that steam automobiles will not be widely accepted for pleasure purposes until the inventors have succeeded in producing a strictly automatic engine.

Much has been said recently as to the use of compressed air for heavy trucks, and several immense corporations have been organized to promote its use. At least one truck has actually been constructed. The air is compressed at a central station, and admitted to heavy steel storage bottles, or tubes, connected with the truck, and is used much like steam. The main difficulty in the process has been the sudden cooling of the machinery when the air is released from pressure and begins to take up heat. Often the pipes and valves are frozen solid. To deal with this problem, a jacket of water heated by a gasoline flame is provided for "reheating" the air, a difficult and cumbersome process. Owing to the weight of the steel tubes, the compressed-air vehicles are enormously heavy, and, like electric vehicles, they must return to some charging-station, after traveling twenty or thirty miles, for a new supply of power. And yet both inventors and financial promoters are sanguine of ultimate success with them.

A Chicago inventor has been building a truck in which he combines gasoline and electrical power. An eight-horse-power gasoline engine situated over the front axle drives an electrical generator, which in turn feeds a small storage battery, thus producing power as the vehicle moves, and rendering it entirely independent of a charging-station. One man can handle the entire truck, and it is said that the cost of operation will not exceed 80 cents a day. The main objection to this system, as with compressed air, is the enormous weight of the vehicle, which is upwards of 9,000 pounds. The truck has a carrying capacity of eight tons, making a total of 25,000 pounds. Such a vehicle presents problems which modern pavement builders have yet to solve.

But the time is certainly coming, and that soon, when all heavy loads must be drawn by automobiles. Recent English experiments, already mentioned, have established the feasibility of the auto-truck even in its present experimental stage, and the inventor needs no further encouragement to prosecute his work. It is hardly possible to conceive the appearance of a crowded wholesale street in the day of the automatic vehicle. In the first place, it will be almost as quiet as a country lane—all the crash of horses' hoofs and the rumble of steel tires will be gone, The vehicles will be fewer and heavier, although much shorter than the present truck and span, so that the streets will appear much less crowded. And with larger loads, more room, and less neces-

An electric hansom cab.

sary attention, more business can be done, and at less expense.

A New York manufacturer produces an odd variation of the motor vehicle in what he calls a "mechanical horse." It is a one or three-wheeled equipment provided with an electric motor, and it can be attached to almost any kind of carriage or wagon and made to draw like a veritable mechanical horse. In this connection, a French manufacturer of a similar equipment says that of the 7,750,000 horse vehicles now used in France, 4,000,000 could be transformed into automobiles, although such a change would probably be impracticable.

Although the American public has not adopted the motor vehicle as rapidly as the French and English, American manufacturers are already well in the lead. It is a significant fact that more vehicles, five times over, are already being exported than are sold here at home. A well-known engineer who has just returned from an exhaustive investigation of automobiles in France says that the European takes an absolutely different view of the automobile from the American.

"The Frenchman," he says, "seems to love his mechanical effects. He makes no attempt to conceal the machinery of his vehicle or to avoid the staggering effect upon the uninitiated of a complex mechanism. His gears are unhoused and his gliding surfaces are left exposed to dust and mud, and he sits among his wheels and levers and brakes and pulleys, a veritable god in the machine. He evidently takes pride in exhibiting his ability to manipulate such a complicated mass of machinery. In America, public enthusiasm has not yet reached the stage in which it can bear the shock of an ordinary examination of such vehicles. We are building carriages, not machines, and making them so simple that a child can run them. Perhaps that is the reason why foreigners are so fond of our vehicles."

As to just what form the future motor vehicle will take there is the widest diversity of opinion. Business clashes with art. Horse carriages are built high so that the driver can see over the horse and avoid the dust. The first motor vehicles were merely "carriages-without-the-horse," and some of them looked

A French touring cart, driven by gasoline.

clumsy and odd enough, "bobbed off in front," as one man described them. Strangely enough, however, manufacturers say that at present the public demands just such vehicles, the low, light, and comfortable models being too much of an innovation to sell.

"But you may depend upon it," one manufacturer told me, "the future motor vehicle will be within a step of the ground, with an artistically rounded front, neither a machine nor a carriage-without-the-horse, but a new and distinct type—the motor vehicle."

The utility of the automobile in any city is in direct proportion to the condition of its streets. It is hardly surprising that manufacturers are receiving the greatest number of inquiries from cities like Buffalo and Detroit, where the pavements are good, and from California and part of New England. The automobile has had such acceptance in France because the highways are all as smooth as park paths. Bicycling already has had a profound influence in spurring the road-makers, and the introduction of the motor vehicle will be still more effective. Colonel Waring estimated that two-thirds of all street dirt is traceable directly to the horse. At present it costs New York nearly $3,000,000 a year to clean its streets. With new pavements such as the new soft-tired vehicles and the absence of pounding hoofs would make possible, street cleaning would become a minor problem. And new asphalt pavement, the best in the world, could be put down at the rate of forty miles a year for what New York now spends for half cleaning its streets.

As yet American law-makers have hardly touched on the subject of motor vehicles. In New York, if drivers keep out of Central Park, display a light, ring a gong, and do not speed faster than eight miles an hour, no one interferes with them. Similar regulations prevail in Boston and in other American cities. In Brooklyn, the parks are free. France and England, on the other hand, hedge in automobile drivers with all manner of rules and regulations, and require them to be officially licensed. In France, by recently promulgated articles, every type of vehicle employed must offer complete conditions of security in its mechanism, its steering-gear, and its brakes. The constructors of automobiles must have the specifications of each type of machine verified by the Service des Mines. After a certificate of such verification has been granted, the constructor is at liberty to manufacture an unlimited number of vehicles. Each vehicle must bear the name of the constructor, an indication of the type of machine, the number of the vehicle in that type, and the name and domicile of its owner. No one may drive an automobile who is not the holder of a certificate of capacity signed by the prefect of the department in which he resides.

The regulations are most explicit on the important question of speed. In narrow or crowded thoroughfares the speed must be reduced to walking pace. In no case may the speed exceed eighteen and one-half miles an hour in the open country or twelve and one-half miles an hour when passing houses. Relative to signals, the regulations say that "the approach of an automobile must, if necessary, be signaled by means of a trumpet." Each automobile must be provided with two lamps, one white, the other green. Racing is allowed, provided an authorization is obtained from the prefect and the mayors are warned. In racing, the speed of eighteen and one-half miles an hour may be exceeded in the open country, but when passing houses, the maximum of twelve and one-half miles must not be exceeded.

One curious difficulty in connection with the new vehicle is the difficulty of finding suitable English names to designate it and its driver. The French, with characteristic readiness in getting settled names for things, have, as already noted, formally adopted the word "automobile" for the vehicle and "chauffeur" (stoker) for the driver. But we of the English tongue are slower. At least a dozen names have been used to a greater or less extent, such as "motor carriage," "auto-carriage," and "horseless carriage." In England, "self-propeller" is popular and so is "auto-car," the latter being apparently the favored designation. Mr. E. P. Ingersoll of the "Horseless Age," who has canvassed the question thoroughly, says that "motor vehicle" seems to be the more generally accepted designation in this country. But whatever it is, or is yet to be, called, the thing itself must now be rated an accepted and established appliances of everyday life. Even if it stopped in its development just where it now is, it must still be accounted of positive and enduring utility; and with the simplifications and cheapening that are sure to be effected by inventive genius and commercial shrewdness in a very short time, its universal adoption is inevitable, and is probably very near.

A Chemical Detective Bureau

The Paris Municipal Laboratory and
What It Does for the Public Health.

By Ida M. Tarbell

July 1894

"**T**HE Municipal Laboratory, " said a physician to me in Paris, "is a chemical police service. Instead of a surveillance over men, it exercises one over compositions. It searches for poisons, microbes, and adulterations, just as the ordinary police searches for assassins, thieves, and embezzlers."

With this remark in mind when I went to see the Municipal Laboratory for myself, I was not surprised to find it installed, as a department of the police, in the Prefecture, a massive pile of buildings facing Notre Dame, and standing in the very heart of Paris. Here it occupies some seventeen rooms in the basement and ground floor. Its present organization dates from 1881, but it really began five years earlier in a station established to detect artificiality in the coloring of wine. The purposes expanded until now the end of the department is to give the people of Paris full information regarding the composition of the food and drink offered for their consumption, and of various other articles (including children's toys and anarchists' bombs) likely to do them harm.

The force employed to prosecute the manifold work of the department consists of a laboratory director, M. Charles Girard, who has been at the head of the institution since the beginning, and may be said, indeed, to have created it; an assistant chief, Monsieur M. Dupres, to whom some of the most ingenious and convenient contrivances peculiar to the laboratory are due; a body of chemists who devote themselves to analysis, each having his specialty of wine, milk, water, or other substance; and a body of expert inspectors, a

Interior of the Paris Municipal Laboratory. From a painting by J.F. Gueldry.

sort of chemical patrol, which, armed with microscopes and endowed with full police power, is free to penetrate into the inner oven of the bakery, the bottommost pit of the grocer's cellar, to take the top layer off every display and look behind every garnish, to confiscate and destroy if it deems best, and to bring back in any case samples of everything suspicious it sees.

The inspectors bring in the larger number of the samples analyzed in the laboratory. In 1889, out of eighteen thousand one hundred and seventeen specimens analyzed, twelve thousand seven hundred and seventy-eight had been submitted by them. But the public is no indifferent patron. Is the coffee muddy, the milk blue, the wine sour, the meat tough, madame or monsieur appears forthwith at the desk where samples are received, to demand an explanation. There is something highly picturesque in the group that gathers in the obscure and rather dingy office. Often there is much that is amusing, so extraordinary are the specimens they submit, the theories they advance. And there is, too, no little that is pathetic. They are not often rich, these people. Most of them wear blouses or black aprons, and rarely is there a woman whose head is covered. Probably there are few the combined wages of whose households average over ten or fifteen francs a day, though the wife, like the husband, has worked her ten hours.

One realizes here, perhaps, as never before, what it means to be poor—that you are the first victim, not alone of epidemic and contagion, but of man's violence and fraud; that because you have not great things, the little that you have shall be taken away. He realizes, too, what such a service may do towards restoring the quality of the poor man's food, and he understands why it is that the proudest boast of M. Girard and his associates is that they have helped to give the Paris working-man better bread and meat and wine.

The name of the article, the date of its receipt, the address of the depositor, and that of the merchant said to have sold it, are noted, and a receipt given the applicant, with directions when to return for the result. The kind of analysis desired is also entered; that is, whether simply a judgment on the quality of the goods presented—the analysis usually asked for by the public and for which there is no charge—or a quantitative analysis, which is a report on the exact chemical constitution. Though the quantitative analysis is less frequent than the qualitative, it yields a revenue not to be despised. In 1889 this amounted to thirty-nine thousand and seventy-five francs.

The laboratories into which the heterogeneous collection of wines and liquors, milk and water, sugar and butter, brass pans and toys, bon-bons and spices, meats and vegetables, firecrackers and dynamite bombs, pickles and canned goods goes, are, in principle, like all laboratories, but still have an air of their own. The scientific operations, too, are those familiar everywhere; yet the direction they take is decided by Parisian habits, and makes the laboratory in a way a reflection of the domestic economy of the city.

Thus no one of the rooms I visited was busier than that devoted to wine. In fact, in 1889, out of eighteen thousand one hundred and seventeen analyses made, six thousand four hundred and fifty were of wines. The proportion is only in keeping with the consumption of the city, which averages about one hundred and fifty million dollars a year, and it is in harmony with the numberless evils which, from the beginning to the end of the life history of a bottle of French wine, combine to ruin its character.

These evils begin with making. Even if the natural process be followed, and the wine made honestly by fermenting fresh grapes, there are various dangerous stages which make manipulations necessary. Suppose that the grapes have been, perforce, gathered before properly ripe. There is an excess of acid in the ferment which must be counteracted, and the sugar must be increased. There are delicate and approved methods for accomplishing this, but they are not always handled skilfully or conscientiously, and some of them give opportunity for a sort of official watering; that is, prescribe a formula which saves the wine and demands enough water to double the vintage.

If the wine escapes in making, it is subject to a multitude of maladies afterwards, which must be treated; and it happens sometimes, as in human medicine, that the remedy is worse than the disease. Litharge, for example, is added to counteract acidity, and is transformed into acetate of lead. Alum is frequently used in diseased wines to give them a certain youthfulness; salt and plaster are standard remedies. But an excess of any one of these substances, or their employment in connection with certain other substances, may result in compounds positively ruinous to the health.

With such manipulations it is only in abusing them, wilfully or ignorantly, that the harm lies. There are others not in themselves harmful, and the

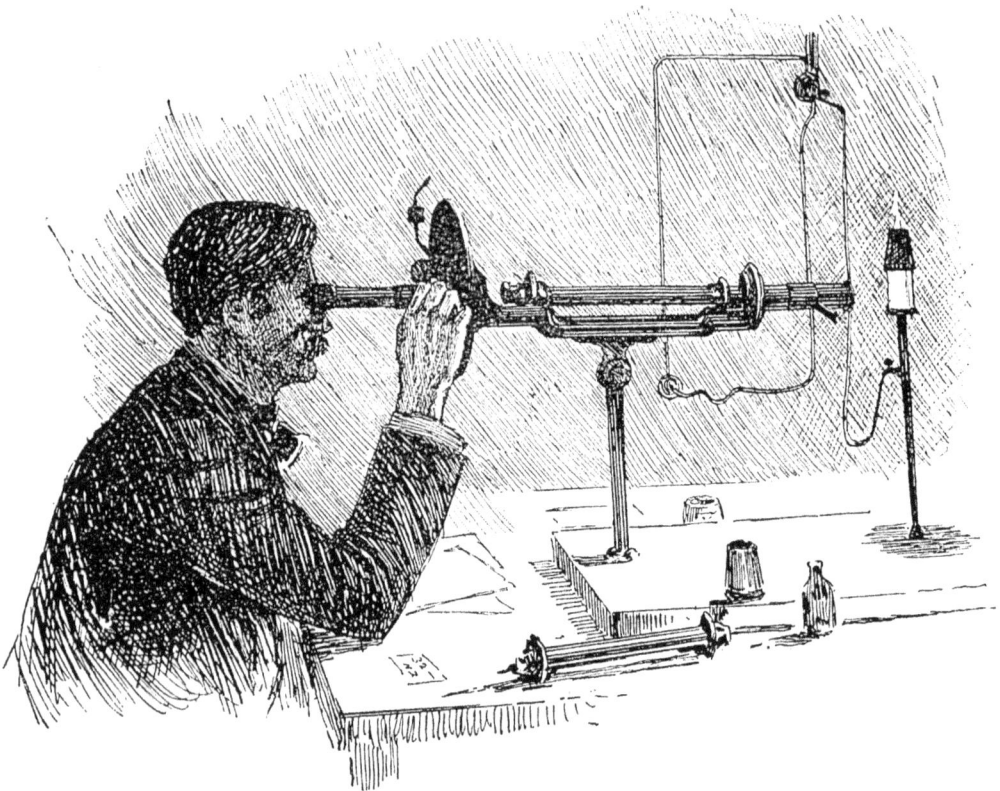

In the dark room. The polarimeter.

chief of them is watering. Thirty years ago this was done in a bold and gross way, simply by adding so much water. It was a fraud, but nobody's health was injured by it. To-day science has come to the aid of the defrauder. Wine weakened by water is strengthened by alcohols of inferior quality, made from grains and beets, producing drunkenness much more quickly than the natural alcohol, and entailing more fatal results. To restore the color lost in watering, various coloring matters, animal and vegetable, are used. The very bouquet is imitated.

But science does still more for the defrauder than this. All of these processes suppose a basis of grape juice. Science has found a way to make wine without this supposed essential, and so perfectly that connoisseurs and chemists hesitate to pronounce it false.

By mixing alcohol, water, saline and coloring matters, and a substance known as the oil of French wine, a composition is produced which many an expert will pass as a natural wine. There is one serious difficulty about this product, however. The oil which furnishes its savor and bouquet is, unhappily, a dangerous poison, a small quantity of which injected into the veins of a dog kills him in less than an hour.

It is the business of the laboratory to decide if any of the manipulations and falsifications hinted at above have been practised on the samples submit-

In the dark room. The microscope.

ted to it. Expert tasters begin the work, and give their judgment on savor, color, bouquet. The chemist then takes it, testing all its fixed and volatile qualities by his sure and delicate processes.

These long and careful examinations give the laboratory the right to speak with decision on the quality of the samples submitted to it. The positiveness of its assertions and its relentless war on defrauders have naturally made it enemies. There are those who complain that the publicity given to the frauds will in the end ruin the foreign wine trade of France. But they have never silenced the laboratory. Its rigor has made the public watchful of what it buys, and more intelligent in the "points " which even an ordinary wine should show. The practice of watering it has greatly decreased.

The public health is not the only gainer. The city treasury gains largely by the decrease of the frauds practised on alcoholic drinks. In 1844 Gay Lussac estimated that Paris lost fully one-third of the *octroi* on wines and liquors because of the falsification which went on inside of the city walls. That it is a matter of importance one realizes when he remembers that of the thirty million dollars in *octroi* which the city put into her pocketbook in 1889, about thirteen million dollars came from the duty oh wines and alcohols. Side by side with the wine analysis in the laboratory are made those on beer, liquors, and ciders. The work is considerable on the first, for the use of beer has made rapid progress in France in recent years. In 1879 only two hundred and ten thousand hectolitres were drunk in Paris. In 1889 *octroi* was paid on 353,122.2 hectolitres. The *brasserie* has become a formidable rival of the *café*. The adulterants of beer, as those of wine, call all the discoveries of science to their aid, and make compounds which for savor, color, and bouquet deceive all ordinary consumers. Malted grains, hops, yeast, and water, the normal materials for producing beer, are all displaced. Glucose or glycerine takes the place of malt. For hops are used beef oils, aloes, quassia, absinthe, gentian, colocynth, salacine, island moss, orange and lemon peel, and various other substances. Alum is used to clear it. The color is improved by caramel, chicory, and various manufactured mixtures.

But it is not only in the making that dangerous compounds are employed. If the dealer fears the beer will not keep, he heats it with such substances as boric and oxalic acid. Nor is this the end of the list of dangers which the laboratory signals in beer. Among the most fatal are the copper, lead, or zinc compounds which it may take up from the vessels in which it has been made, or from the pipes and faucets of the casks and reservoirs in which it is kept.

Though alcoholic drinks are in excess in the laboratory, I found that they did not absorb it. Milk, "the wine of the children, " has been since the begin-

Apparatus for testing water.

ning one of its chief objects of investigation. In 1881, when the investigations of the milk supply of Paris began, 50.6 per cent. of the samples analyzed were "bad." In a year, thanks to the vigor of the service, this percentage was reduced to 30.7. In 1889 10.6 per cent. of impure were found on three thousand seven hundred and ninety-five analyses.

I was curious to get the judgment of the laboratory on the Paris water, for I had been remonstrated with persistently for drinking it. I applied to one of the chemists in the department devoted to water analyses, who, for reply, took out several bottles containing waters of the various kinds used, and named according to their source, water of the Vanne, Dhuis and Arve.

"These waters, " he said, "have stood here a week. They are absolutely pure, answering to the laboratory standard of wholesome water in all particulars. A city could not have a more satisfactory water supply than we have now. It is true that it is only since last spring that there has been enough that is pure for the entire city. In the environs, the water is positively dangerous."

This water department of the Municipal Laboratory has rendered another service to Paris in showing the danger of using the wells, of which there have been a great number within the walls. The well water is heavily impregnated with lime, and the Paris bakers claim that a sponge mixed with it is much lighter and better than that mixed with hydrant water. If in baking all possible germs were killed, there would be no danger in allowing the practice, but the laboratory has found that the heat of the oven is not sufficient. Typhus and cholera microbes both might pass through a baking unharmed. When this was established, the inspectors began a vigorous campaign against the wells, destroying them from the foundations.

The importance of thus superintending the bakers is evident when it is remembered that in Paris no one bakes at home, and that bread and pastry are always bought. Nothing could be more inviting than the public *boulangeries* and *patisseries*. They are models of neatness, good taste, and tempting displays. But "things are not always what they seem, " and there are more points to oversee in these shops than the water with which the bread is mixed.

The flour gives the Municipal Laboratory no little trouble. It is found mixed with sand, chalk, plaster, alum, phosphate of lime, carbonate of magnesium, even sulphate of copper. Wheat flour is adulterated with cheaper kinds, as rye, barley, corn, pease, beans, lentils, rice, millet, buckwheat, potatoes, even with fine sawdust. Then butter is replaced by oleomargarine; the sugar with that "sweetened illusion " saccharine. If the cakes escape the adulterated flour, butter, and sugar, they still run the chance of being colored with some injurious substance. A cake is a work of fantasy in Paris. It imitates everything created or manufactured, from a canary bird to a Swiss chalet. To carry out such ambitious designs, colors must be employed, and frequently they are poisonous, though the laboratory has published a careful list of what materials can be employed safely in coloring sugars and bon-bons.

The inspection of the markets is an especially interesting part of the service, for the cleverest devices are practised in disguising tainted fruits, vegetables, and meats, and in keeping the attention of the inspector away from the weakest spot in the stock. The agent must match address with address. In case fraud is found in any of the perishable articles, it is confiscated on the spot.

A great deal of half-spoiled merchandise is found in the carts of the wandering merchants. The confiscation of the stock is almost always a sad busi-

ness. Women and decrepit men form the body of this band of merchants, and the loss often must take away the bread from them and their children.

In the survey of the butcher shops, one duty is to make sure that horse, ass, and mule meats are not masquerading as beef. Not that their sale is forbidden. On the contrary, the Municipal Laboratory itself has declared this sort of nourishment "an excellent thing. " It simply demands that the meat be sold as equine and not bovine, and that the animals which furnish it be not decrepit or diseased.

The first point is regulated by establishing shops especially for the sale of horse meat. Or, if it is sold from a cart or in a regularly licensed butcher shop, by requiring that it be marked plainly. The sale of horse meat has grown to enormous proportions since the first shop was established in 1866. The estimate is, that it is eaten now in a third of the Parisian households. In 1891 twenty-one thousand two hundred and thirty-one horses, sixty-one mules, and two hundred and seventy-five asses were sold in the Paris shops. The meat costs about half as much as beef.

The inspectors find a great deal to do in the groceries. The adulteration is particularly common in spices, tea, coffee, chocolate, and canned goods.

I was particularly struck by the number of cooking utensils I saw heaped up in one of the rooms at the laboratory. "Confiscations of the inspector, " said the chemist.

"Do you survey kitchens, then?"

"Certainly, " he responded. "Every dish used in a public restaurant of Paris, either in the kitchen or for the table; every pot, pan, and utensil in the bakeries; and every beer faucet in the wine shop—in short, everything used in preparing or serving foods, is under the care of the inspector. The law forbids the use of lead, zinc, and galvanized iron in the manufacture of cooking vessels. It orders that all copper vessels be tinned and kept in good condition. It directs that pottery which is covered with a glaze containing enough oxide of lead to yield to a feeble acid be seized. It orders that tin cans never be soldered on the inside, and that the materials used in their manufacture be conformed to a certain standard. It is the inspector's business to look after all these things."

"And the results?"

"That depends. There are establishments in Paris, like the great restaurants, which employ a skilled tinner regularly, and their utensils are always in order. In many little shops kept by women the copper vessels are the pride of the establishment, but in many others they are, unhappily, neglected. In

1889, out of two hundred and fifteen samples analyzed here, ninety-seven contained lead."

We had reached the office where the samples are handed in. A woman was at the desk with a bit of cheap colored candy, which one of the service was examining. "It is no doubt this stuff, " he was saying, "which has made your baby sick. You must not buy colored bon-bons."

"There," said my guide, "is one of the reassuring parts of our service. That woman will receive to-day a lesson she will never forget. All her neighbors will hear it from her, and it will probably become a tradition in her family that colored sweets are dangerous. Very often, too, we give them simple methods for detecting frauds. Thus they become their own inspectors."

"Do they often prosecute the dealers? " I asked, as the woman vented her wrath in a torrent of invectives against the merchant who would sell poisons, threatening him with arrest and imprisonment.

"No. We advise them not to. For they rarely have proof to show that the sample came from the merchant charged with selling it, and they not only lose the case, but pay expenses. Their best plan is to change merchants. As they are obliged to leave the address of the person charged with selling the goods, our inspectors examine his stock, and if the samples taken are bad, they give the court the information, which enables it to punish him promptly."

"But the merchant may be deceived?"

"Of course. But it is his business to know the quality of what he sells. To aid him we publish full reports of processes for detecting frauds, and the laboratory is as free to him as to the retail buyers. Indeed, we urge merchants to submit samples before giving large orders, and we have arranged it so that they need not go to the trouble of bringing them here, but may leave them at the police station of their quarter, and we return them the analyses. As a rule, of course, the dealer regards us as his enemy; but we are his friend if he is an honest man, protecting him from the manufacturer."

"But the manufacturer may be ignorant of the dangerous quality of the materials he employs."

"According to French law such ignorance is a crime, and we do our best to inform him. Take the matter of coloring toys, for example. We publish a full list of dangerous coloring substances. We show the poisonous compound which may result from the use of certain materials in wines and beers. We show the effects of every suspected method, of every suspicious element, in the preparation of foods and drinks and the manufacture of articles. So if the manufacturer knows his business, he need not produce goods unfit for the market."

"And you hope some day, I suppose, to prevent his doing so; to know for a certainty that nothing tainted, adulterated, or watered is sold in Paris?"

He smiled cynically. "When the Prefecture of Police is unnecessary, the Municipal Laboratory will be also; but not before."

Investigations

Scientific Kite-Flying

By Cleveland Moffett

March 1896

O N the long peninsula that separates New York Bay from Newark Bay, there is, among other things, a red house by an open field, in which lives the king of kite-flyers. Every one in Bayonne, the town which covers this peninsula, knows the red house by the open field; for scarcely a day passes, winter or summer, that kites are not seen sailing above this spot—sometimes a solitary "hurricane flyer," when the wind is sweeping in strong from the ocean; sometimes a tandem string of seven or eight six-footers, each one fastened to the main line by its separate cord. And wonderful are the feats in kite-illumination accomplished by Mr. Eddy (the king aforesaid) on holiday nights, especially on the Fourth of July, when he keeps the sky ablaze with gracefully waving meteors, to the profound awe or admiration of his fellow-townsmen.

If you enter the red house and show a proper interest in the subject, Mr. Eddy will take you up to his kite-room, where sky-flyers of all sorts, sizes, and materials range the walls—from the tiniest, made of tissue paper, to nine-footers, with lath frames and oil-cloth coverings. Hanging from the ceiling is one of the queer Hargrave kites, which looks like a double box, and seems

Hargrave lifted sixteen feet from the ground by a tandem of his box-kites.

as little likely to fly as a full-legged dining-table; yet fly it will, and beautifully too, though by a principle of aëroplanes only recently understood.

Then Mr. Eddy will show you the room where, with the help of his deft-fingered wife, also a kite enthusiast, he spends many hours developing and mounting photographs taken from high altitudes, with a camera especially constructed to be swung and operated from the kite cord.

Until one talks with a man like Mr. Eddy—though, indeed, there is no one just like him—one does not realize what a large and important subject this of scientific kite-flying is. Many men of distinction have devoted years of their best energies to experiments with kites. Mr. Eddy himself is a scientist first, last, and always; for the sake of a new observation he will send up a tandem of kites when the thermometer is below zero, or stand half a night at his reeling apparatus, getting records of the thermograph.

Perhaps I shall do best to begin by giving some useful information to those who may contemplate constructing a modern scientific kite. The first thing that should be done by such a person, be he boy or man, is to rid his mind of all his preconceived notions about kites, for it is almost certain that they are incorrect. To begin with, the scientific kite has no tail. A few years ago people would have laughed at any one who attempted to send up a kite without a tail. But the question is now no longer even open with the scientific kite-flyers, who not only send up tailless kites with the greatest ease, but do so under conditions which, to kites with tails, would be impossible: for instance, in dead calms and in driving hurricanes. The tailless kite, sent from the hands of a master, will fly in all winds.

It is true that kites with tails have given good results in experimental work; but the tails are annoying and an unnecessary weight, and may better be dispensed with. Every boy has had the vexatious experience of sending up a kite in a light breeze with a tail made light in proportion, only to find that, on reaching stronger air currents above, the kite has begun to dive and grow unmanageable. Then, when he has taken the kite down and added a heavier tail, he has found the breeze at the ground insufficient to lift the extra load; and so, between two difficulties, has had to give up his sport in disgust. This is the one serious defect of kites with tails, that they cannot adapt themselves to wind currents of varying intensities; whereas the tailless kites do so without difficulty. And in tandem flying, which is the backbone of the modern system, the weight of a half dozen or more heavy tails would be a serious impediment, to say nothing of the perpetual danger of the different tails getting entangled in the lines.

HOW TO MAKE A SCIENTIFIC KITE.

It is important, then, to know how to make a scientific tailless kite, such as is used by the experts at the Smithsonian Institution, or at the Blue Hills Conservatory near Boston, for it must not be supposed that kite-flying is merely an idle pastime; it is a pleasure doubtless for boys, but it is also a field of serious experiment and observation for men. The information I here present, including practical directions as well as interesting theories, was obtained from Mr. Eddy himself, and may be regarded as strictly accurate.

It is much better for amateurs to begin with a kite designed to fly in strong winds, as it is a long and delicate task to learn to manage the variety with extra wide cross-stick meant for ascension in calms. The two sticks which form the skeleton should be of equal lengths, say six feet; and should cross each other at right angles at a point on the upright stick eighteen per cent, of its length below the top. This point of crossing is of great importance, and was only located by Mr. Eddy after months of wearisome experiment. He was misled in his earlier efforts at tailless kite-making by the example of the Malay kiter-flyers, who are reputed to be the most skilful in the world, and who cross the sticks much nearer the middle of the upright one. In a six-foot kite the two sticks, equal in length, should cross at about thirteen inches from the top of the upright stick; and the same proportion should be observed for kites of other dimensions. At the point of crossing, the sticks should be slightly notched, and strongly bound

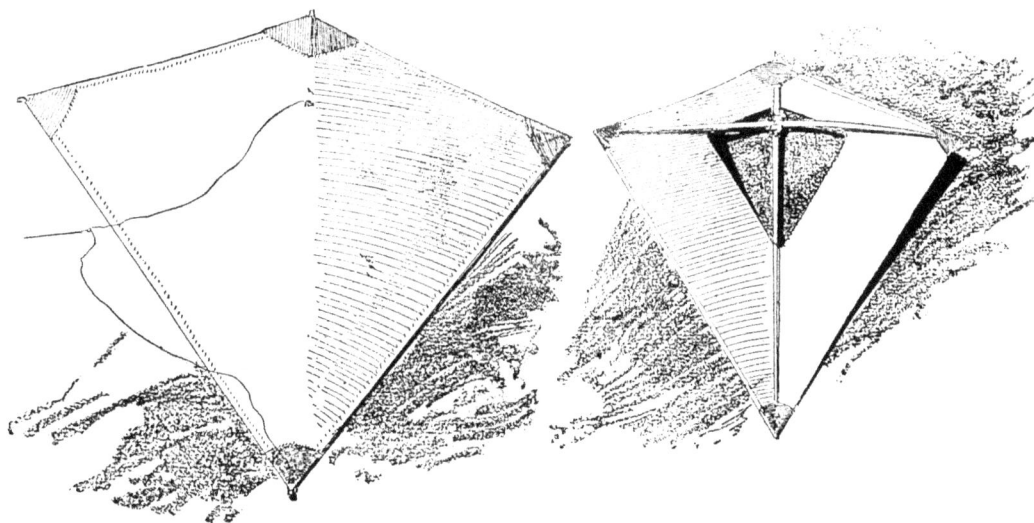

The Eddy Tailless Kite

Left: Front view, showing how the line is attached. Right: A storm-flyer.—The diamond-shaped figure in the centre is an opening made to lessen the wind pressure.

together with twine tied in flat knots. Driving a nail or screw through the sticks, to bind them, weakens the frame at the point of greatest strain.

As material for the sticks Mr. Eddy has found clear spruce better than any other wood. Bamboo is bad, because it bends unevenly at the joints. White pine is not tough enough, and cypress is both too brittle and too flexible. The hard woods, like ash, hickory, and oak, are too heavy; in scientific kite-flying, even so small a weight as a quarter of an ounce may make all the difference between failure and success. All winds are broken by frequent brief intervals of calm, and a kite must rely on its lightness to outride these. Whoever contemplates going seriously into kite-flying will do well to provide himself with a store of suitable sticks by purchasing a straight-grained, well-planed spruce plank, free from knots, and having it sawed on a circular saw into sticks five-sixteenths and seven-sixteenths inches in thickness, to be cut later into such lengths as he may choose.

The two sticks (there are never more than two) having been fastened firmly together, the cross-stick must be sprung backward; so that, when finished, the kite will present a convex or bulging surface to the wind. It might be imagined that a concave surface to the wind would be better; and indeed this has been tried. But it has invariably proved that with a concave surface the kite receives too much of the breeze and becomes quite uncontrollable. The amount of spring that must be given the cross-piece is in proportion to its length, Mr. Eddy's rule being to spring the cross-stick, by means of a cord joining the two ends like a bow, until the perpendicular between the point of juncture of the two sticks and the centre of the cord is equal to one-tenth of the length of the cross-stick, or a little more than one-tenth, if the kite is to be flown in very high winds.

It is of the first importance o keep the two halves of the kite on the right and the left of the upright stick perfectly symmetrical. And this is by no means an easy matter. It often happens in bending the cross-stick that, owing to differences in the fibre and elasticity of the wood, one side bends more than the other, with the result that the two halves present different curves and consequently unequal wind areas. To offset this difficulty, and also to strengthen the skeleton, Mr. Eddy's practice is to add a bracing piece at the back of the cross-stick—a piece about one-fourth of the length of the cross-stick itself, and of the same width and thickness. If the two halves of the kite are already quite symmetrical, he places this bracing stick with its centre directly even with the point of juncture of the two large sticks, its two ends being fastened with twine to the cross-stick, about nine inches on either side of the crossing-point. But if one half of the cross-stick shows a greater bend than the other, he places the longer arm of the bracing piece toward the side that bends the most, thus presenting a greater leverage against the wind on that side than on the other, and so equalizing things.

With the two sticks and the brace all thus properly in place, a supporting frame for the paper or cloth is formed by running, not cord, but fine picture wire, over the tips of the sticks, notched to hold it in place, in the ordinary way. Then, with a thin, clear paste made of starch, the paper may be laid on, care being taken to paste the edges so as to leave a certain amount of slack or looseness in the part of the kite below the cross-stick, so that each of the lower faces will present concave wind surfaces. To preserve the required equilibrium, it is important that the amount of looseness in the paper be equal on the two sides; and in order to keep it so, it is necessary to measure exactly the amount allowed.

Those who wish to make many kites will do well to buy thin manilla paper, as wide as possible, having the dealer roll off for them seven hundred or eight hundred feet, say a yard in width, which will insure a cheap as well as an abundant supply. For strong winds and large kites it is best to use cloth as the covering. It should be sewed to the frame, and, if carefully put on, will do service for years. Silk, of course, is the ideal material; but its costliness puts it beyond ordinary means, and common silesia, such as is used in dress linings, is almost as good. Whatever the material, the kite should be fortified at the corners by pasting or sewing on quadrants of paper or cloth, so as to give double thickness at the points most liable to injury. A finished six-footer should not weigh over twenty ounces, if covered with paper; or twenty-five ounces, if covered with cloth. Mr. Eddy has made a six-footer for calm flying as light as eight ounces.

HOW TO SEND UP A KITE.

There is only one way to learn the practical art of kite-flying, and that is to begin and do the thing yourself—with many mishaps and disappointments at the outset. One of Mr. Eddy's practices when sending kites up in very light winds or in an apparent calm, is to reel out two hundred yards or so of cord in a convenient open space, leaving kite and cord on the ground until ready to start. Then, by taking the cord at the extreme distance from the kite, and beginning to run with it, he gets it quickly into the upper air currents, which are always stirring more than those at the surface. It is sometimes necessary to run for a considerable distance before the kite reaches a sustaining current;

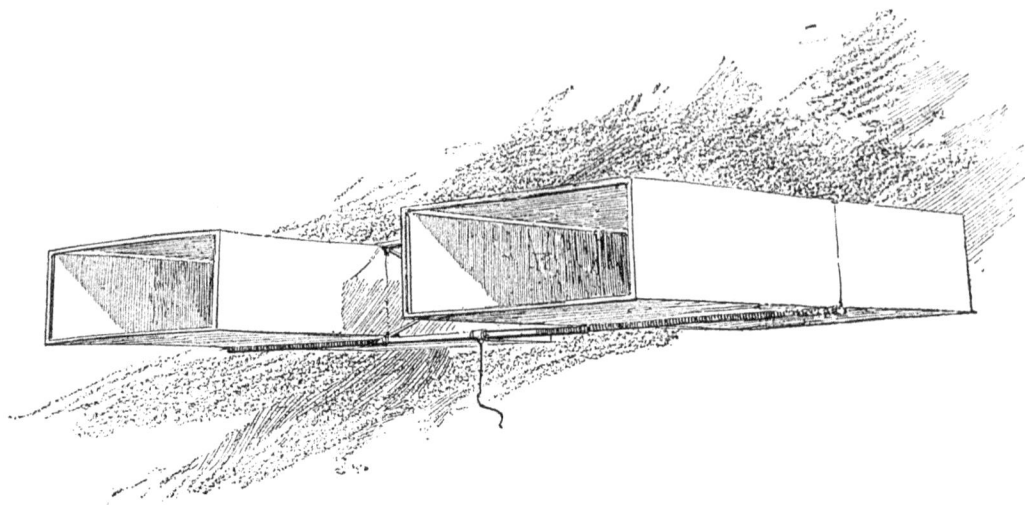

The Hargrave Box-kite

It was by kites of this variety, flown in tandem, that the inventor, Hargrave, was lifted sixteen feet from the ground on November 12, 1894.

but a real kite enthusiast will not mind taking trouble; indeed he had better abandon the whole business if he does. It is worth noting that even in a dead calm a kite may be kept up indefinitely as long as the flyer is willing to run with the cord at the rate of about five miles an hour.

In flying kites tandem there is always to be guarded against the danger of a breaking of the cord. Few people realize how hard a pull is exerted by a series of kites well up in the air. A strain of twenty-five or thirty pounds on the cord is not uncommon; and not only the strength of the cord, but the way of attaching it, is of great importance. There should be two strings (never more), fastened to the upright stick at its lower end and at the point of crossing, the upper length being about one-third of the lower one, and the two being adjusted so that, when taut, the kite takes an angle of about twenty degrees with the ground—which means that the kite goes up almost straight overhead, the string making an angle of about seventy degrees with the ground.

In sending up a series of kites to fly tandem, it is best to head the line with a small kite, three or four feet in diameter, and gradually increase the size until a diameter of six feet is reached for the one sent last. This arrangement makes it possible to hold the upper kites by lighter cord, the heavier kites being reserved for the half of the line nearest to the ground; and thus there is a material lessening of the load to be borne. The first kite should be well up, say five hundred feet, before the second is attached to the line. But after that they maybe sent at closer intervals, sometimes with only a few hundred feet between them—say two hundred feet in light winds, and five hundred feet in heavy winds. Each kite in a tandem should have a length of at least one hundred feet of cord from the main line, and great care should be exercised in knotting fast the individual lines.

The best way of starting a second kite, after the first is well up, is to pay out about a hundred feet of cord for the tandem line, attaching one end of this to the main cord and the other to the second kite, which is left lying on the ground back downward. Then pay out the main line evenly until the tandem line begins to lift. As the pendent kite is borne higher and higher, it will swing for a while in a horizontal position; but will presently begin to flutter and sail sideways, and them finally come up more and more, until the wind catches it and it shoots up like a bird into its proper position. In fact, once the first kite is securely up, the others will fly themselves by merely being attached to the main line as described. Of course each fresh kite increases the pull on the main line, and the line must be made proportionately stronger as the tandem is increased.

RUNAWAY TANDEMS.

Mr. Eddy has had some remarkable experiences with escaping kites. One day at Bayonne, in July, 1894, while he was flying a tandem of eight kites in a northwest wind blowing eighteen miles an hour, the main line broke with a loud snap, and the kites sailed away towards Staten Island with the speed of an escaped balloon. One can scarcely conceive the rapidity with which a line of kites like this travels over the first four or five hundred feet after its release. An ice-boat goes no faster, and one might as well pursue the shadow of a flying cloud as chase that string. At the time of the escape the top kite, a four-footer, was up nearly a mile, and the other seven were flying at a good elevation. The consequence was that although, as invariably happens in such cases, they began to drop, the lowest kite did not strike the ground until it had been carried about a quarter of a mile, to the New Jersey shore of the Kill von Kull, which is half a mile wide at this point. Here kite number eight, a six-footer, caught in a tree and held the line for a few seconds until its own cord broke, under the strain, and set the other kites free. This check had lifted the other kites, and they now flew right bravely across the water, not one of the seven wetting its heels before the farther shore was reached. Then the lowest of them came to the ground, in its turn putting a brief check on the others. But its cord soon broke under the strain, and the six still flying went sailing over the trees of Staten Island, hundreds of people watching them as they flew—

six tailless kites driving along towards New York Bay, the main line trailing behind over lawns and housetops.

Then a queer thing happened. As the loose end of the main line trailed along, it whipped against a line of telegraph wires with such violence as to wind itself around the wires again and again, just as a whip-lash winds round a hitching-post when whipped against one. The result was that the runaway kites were finally anchored by the main line, and held fast until their owner, coming in quick pursuit on ferryboat and train, could secure them.

On another occasion, two of Mr. Eddy's kites flying in tandem broke away, and started out to sea, the dangling line passing over a moored coal barge on which a man was working. Feeling something tickle his neck, the man put up his hand quickly and touched the kite-cord. Greatly surprised, he seized the cord and made it fast; and he was not at all disposed to give up the kites when Mr. Eddy claimed them. There is no property, indeed, so hard to prove and recover as a runaway kite. For one thing, there is absolutely no telling how far a runaway kite will sail before landing. Mr. Eddy estimates that when the main line breaks, a kite well up in a twenty-five mile breeze will travel, before alighting, a distance equal to twelve times its height from the ground. This means that a kite straight over the Battery, in New York City, and a mile in the air, driven by a stiff south wind, might land in Yonkers if the cord broke. There is, by the way, an old-time ordinance on the statute book, prohibiting the flying of kites in any part of New York City below Fourteenth Street. This, however, did not prevent Mr. Eddy from taking recently a series of unique photographs (some of them are reproduced in this article), by means of a tandem of kites sent up from a high building near the City Hall Park. The only complication that resulted was a fierce contention among a crowd of idlers and gamins over the possession of one of the kites, which came down accidentally and lodged in one of the Park trees.

THE LIFTING POWER OF KITES.

A tandem of six or eight six-foot kites exerts a pull of thirty pounds or more on the main line; but it must not be assumed that such a tandem would lift and carry through the air a weight of thirty pounds. The weight of thirty pounds would be carried a short distance; but as the weight moved off, there would be a sudden lessening of the resistance on the line, and so of the wind pressure against the kites, which would soon cause them to sink. A tandem of strong kites in a good breeze might be made to operate a sort of jumping apparatus which, after being carried a short distance, would anchor itself to the ground until the renewed strength of the kites lifted it up again for another jump. But all kite experts are agreed that a kite's power for lifting loads clear of the ground must be enormously increased according as the distance to which the load is to be lifted is increased. It would be possible, for example, to build a tandem of kites strong enough to lift a man clear of the ground, supposing him to be swung in a basket from the main line. This, indeed, has been actually accomplished. September 18, 1895, in England, Captain Baden-Powell was lifted to a height of one hundred feet on a kite-string supported by five large hexagon kites. But Mr. Eddy calculates that to lift a man of the same weight (one hundred and fifty pounds) to a height of fifteen hundred feet, with a wind blowing at the same rate (twenty miles an hour), would require seven kites with upright and cross-sticks not less than sixty-four feet each in length.

The only other instance on record where a man has been lifted by a kite-cord was in the experiment of the great Australian kite expert, Hargrave, who, on November 12, 1894, placed himself in a sling seat attached to a tandem of his wonderful box kites, and was swung sixteen feet clear of the earth. The entire load, including the seat and appurtenances, amounted to two hundred and eight pounds. Mr. Eddy calculates that six of his bird-shaped kites, twenty feet in diameter, would lift a man and basket in safety to a height of one hundred feet, assuming the wind to be blowing steadily at twenty miles an hour.

THE METEOROLOGICAL USE OF KITES.

Although Mr. Eddy began flying kites as a diversion, he soon saw that there were more serious reasons for continuing his experiments. Having long been interested in meteorological problems, it occurred to him that good results might be obtained by sending aloft, on kite-strings, self-registering thermometers and apparatus for indicating the direction and strength of the air currents. On February 4, 1891, he sent up what is believed to be the first thermometer ever attached to a kite for scientific purposes. This was at nine o'clock in the evening on a cold winter's night, the thermometer registering ten degrees Fahrenheit at the ground. On reading the record after the descent, the thermometer was found to mark six degrees Fahrenheit, which indicated, according to the recognized law of decrease of temperature, that the kite had been sent to a height of one thousand feet. The law is that in ascending from the earth the temperature falls one degree for every two hundred and fifty feet; but subsequent experiments convinced Mr. Eddy that it was by no means to be relied upon as an indication of the height of kites. Not that the law is false; but it holds good only when the meteorological conditions above are the same as at the earth's surface, which is very far from being the case always.

Out of these experiments Mr. Eddy evolved an important theory which has since been abundantly verified. Seeing the frequent variations in the thermometric readings from what the law had led him to expect, he concluded that these were due to meteorological variations overhead; and that changes in the weather, say the approach of warm waves or cold waves, make them

New York, East River, Brooklyn, and New York Bay from a kite.

From a photograph taken from a kite by Mr. W. A. Eddy.

selves felt in the air strata above the earth's surface several hours before they can be detected at the surface. Observations extending over months at the Blue Hills Observatory, near Boston, and elsewhere, have abundantly confirmed this theory.

With this fact established, it followed, in Mr. Eddy's opinion, that it was perfectly possible to use kites in making weather prognostications; and, indeed, he has been doing this himself for several years with the best results. Whenever his kite-thermometers, sent to a fixed height which he determines independently by a specially devised kite-quadrant, show actual readings which are either warmer or cooler than the theoretical readings, he prophesies that the weather will, within a few hours, become warmer or colder at the earth's surface, and these prophecies are fulfilled in a large majority of cases. If the kite-thermometers show exactly the temperature which the law would call for, he prophesies that there will be no change in the weather.

It has also been demonstrated that kites may be used by meteorologists to indicate the approach of storms, which they foretell by a sudden and continuous veering over a considerable arc, usually about sixty degrees. This veering begins usually six or seven hours before a storm, and often as much as twelve hours. And another sure sign of a storm is the continuous and sudden dropping of the kites followed by a quick recovery, which shows that the wind is blowing in gusts interspersed with periods of calm.

In making a series of meteorological experiments which he conducted at the Blue Hills Observatory, Mr. Eddy often employed as many as eight or ten kites; and in August, 1895, he sent up twelve kites on one line, three of them being nine-footers. This is probably the largest number of kites ever sent up in tandem; and although on this occasion the line carried only the thermographs suspended in a basket, the whole weighing not more than two pounds, a very much larger load might have been carried, had it been desired.

Among many other curious things about the wind observed by Mr. Eddy, is the fact that the night winds are by far the steadiest and most satisfactory for kite-flying. On this account much of his work with kites has been done in the darkness, although he uses lanterns on the lines to assist him in locating the kites. It has also been demonstrated that the force of the wind increases steadily as the distance from the earth increases. Archibald proved this conclusively, by suspending a series of wind-measuring instruments at intervals along the main line, their registration showing almost invariably greater wind pressure at the higher altitude. Mr. Eddy has furthermore noted that, while the early morning wind is usually very light at the earth's surface, it is almost invariably good aloft; and he has again and again verified the well-established fact that all clouds herald their approach and are accompanied by increased wind velocity.

THE HIGHEST FLIGHT EVER MADE BY A KITE.

The modern system of flying kites tandem was devised by Mr. Eddy in 1890, although it was hit upon two years later independently by Dr. Alexander B. Johnson, the distinguished surgeon of the Roosevelt Hospital in New York. The tandem system makes it possible to send kites to far greater altitudes than had ever been previously attained. And here the best record is undoubtedly held by one of Mr. Eddy's tandems, sent aloft at Bayonne, on November 7, 1893. Mr. Eddy began to send up the kites at 7:30 A.M.; but, being hampered by light breezes from the east, found he was kept busy until half-past three in the afternoon in getting nine kites aloft. He had paid out nearly two miles of cord, when the top kite, a little two-footer, stood straight over the spar buoy in Newark Bay. The lowest kite, a six-footer, was hovering some distance inland from the shore, on a line from the shore to Mr. Eddy's house (where the end of the line was anchored) measuring fifty-five hundred feet by the surveyor's map. Taking two observations from the two ends of this base line, Mr. Eddy's kite-quadrant showed angles of thirty-five and sixty-six degrees; and these data, by simple methods of triangulation, were sufficient to determine the altitude of the kite, which was found to be five thousand five hundred and ninety-five feet—or something over one mile. The kites were seen by hundreds of persons during the fifteen hours that they remained up, the experiment coming to an abrupt end at ten o'clock that night by the blowing away of the two upper kites in the increasing wind. The escaped kites disappeared in Newark Bay, along with three thousand feet of the line.

Much interest attaches from a scientific point of view to experiments designed to test how great an altitude may be reached by kites; and for a year past Mr. Eddy has been working in this direction for the Smithsonian Institution, the hope being that he will ultimately succeed in sending kites two miles above the earth's surface. Professor Langley has been following these experiments with great interest, and has furnished Mr. Eddy with a special quality of silk cord which, it is believed, will give better results in meteorological observation than the ordinary hempen twine or rope. The great difficulty that Mr. Eddy finds in the way of making his kites reach great altitudes, is the pull on the cord, which increases greatly as the kites rise higher. It is probable that a tandem of fifteen or

Photographing from a kite-line.

Note.—In this picture the square box suspended from the upper line is the camera. The ball hanging from the camera is the burnished signal which, by its fall, informs the operalor on the ground when the shutter of the camera has opened. The shutter at the ball are controlled from the ground by the lower line.

twenty big kites, reaching to a mile above the earth's surface, would exert a pull of one hundred pounds; while at a height of two miles they might, Mr. Eddy thinks, exert a pull of three hundred and fifty pounds; and at a height of three miles, a pull of seven hundred pounds. However great the pull, it is essential to successful flying that the man in control be able to let out or reel in the main line with great rapidity, and it is evident that a dozen men could not by hand alone accomplish this if the kites were sent as high as might be. It is likely, therefore, that, as the importance of scientific kite-flying becomes more widely understood, some simple dummy engine will be devised for rapidly turning the windlass on which the main line is wound.

Mr. Eddy has made frequent experiments with rain-kites, which he used for the first time in November, 1893. It is true that Franklin sent up a flyer during a shower, but in his case the rain was merely an accident accompanying the electric storm, which was his only concern. Mr. Eddy, however, has sent up kites in the rain for the purpose of studying cloud altitudes and other meteorological phenomena; and by this means he has discovered what was not previously believed to be true: that clouds sometimes sink to within six hundred feet of the earth's surface without actually coming down to it. In fact, Mr. Eddy has had kites disappear in a cloud at a height of only five hundred and sixty-eight feet. It has sometimes happened that clouds settling toward the earth have obscured the kites gradually, the top one becoming invisible first, and then the others in succession. Mr. Eddy has found that by such indications he is able to foretell the approach of fog four or five hours before it reaches the earth's surface, so slowly do the clouds settle through the air strata.

It is best to make rain-kites of oil-skin or paraffine paper, as the ordinary paper or cloth becomes saturated with the dampness and very heavy, thus lessening the buoyancy of the line. So penetrating is the dampness of clouds, even without a rain-storm, that the wooden frames sometimes become warped and the paste seams soak open.

DRAWING DOWN ELECTRICITY BY A KITE-STRING.

The scientific kite-flyer will find much to tempt him into the field of electricity; and will be able, not only to duplicate Dr. Franklin's historic experiment of bringing down sparks from the heavens, but may go far beyond this, taking advantage of the greater knowledge of electricity at his disposal and the superior apparatus. In the summer of 1885, Alexander McAdie, at the Blue Hills Observatory, got strong sparks at the earth's surface from a wire

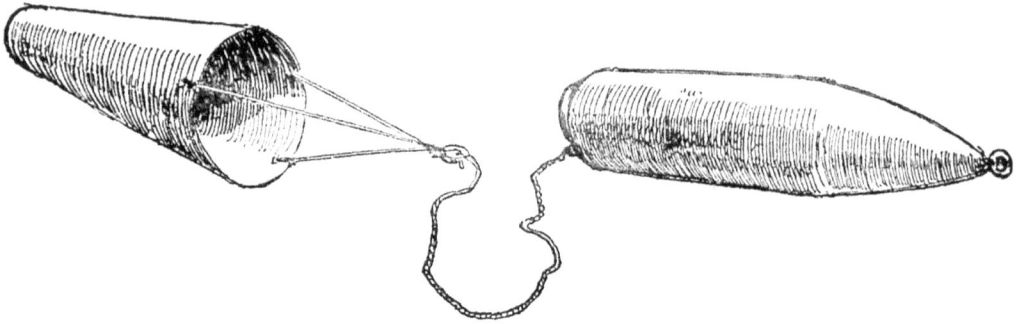

Kite-drawn buoy.

Invented by Prof. J. Woodbridge Davis. This buoy lacks the steering appliances of the one shown below, and travels simply in a line with the kite that draws it.

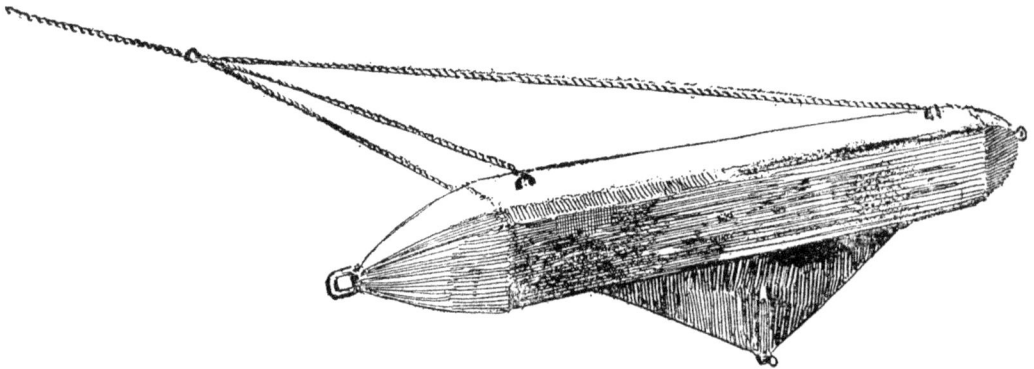

Dirigible kite-drawn buoy.

This is the buoy invented by Prof. J. Woodbridge Davis for conveying messages, food, or life-lines between disabled vessels and the shore. The buoy is drawn over the water by the kite-line, like the one shown above, but the setting of the keel and the three guy-ropes give it whatever direction is desired.

connected with a kite whose surface had been coated with tinfoil so as to form an electric collector. He also, by the brightness and increased lengths of the sparks obtained, proved that the electric force in the atmosphere is very greatly increased with the approach of thunder clouds; and also that this force increases steadily as the kites reach greater altitude, and *vice versa*. Indeed Mr. Eddy and others who have conducted similar experiments, have found the electric force so strong at certain altitudes as to make the manipulation of the conducting wire a source of considerable danger.

On October 8, 1892, Mr. Eddy made an important advance in electrical experiments with kites, by using a collector quite separate from the kites themselves, which were merely used in tandem to support the line on which the collector was swung and raised to any desired altitude. By this arrangement any accident that might befall one of the kites is less likely to ruin the whole experiment.

Much experience with the kite-collector has convinced Mr. Eddy that there is always in the air overhead, at all times of the year and in all weathers, an abundant, practically a boundless, supply of electricity. It has never yet happened to him to send his collector up to even so low a height as four hundred feet without getting a spark in his discharge-box at the earth. He has discovered, however, that the greater the amount of moisture in the air, the greater is the height to which he must send the collector before getting the first spark. There is no doubt that large quantities of electricity might be obtained by hoisting large collectors, supported by strong flying tandems, to considerable altitudes, and drawing off the supply at the earth by means of a system of transformers which would lower the electricity from the dangerously high tension at which it discharges down the wire, to a voltage that could be handled with safety. In his experiments thus far, Mr. Eddy has discharged the copper wire leading from his collector into a wooden box containing a pasteboard wheel with darning-needle axle and tinfoil edges. The axle is grounded, and the copper wire from the collector placed near the tinfoil periphery of the wheel, so as to discharge its sparks through the intervening distance, and by the shock cause the wheel to turn.

THE USE OF KITES IN PHOTOGRAPHY.

One of the most interesting applications of the kite, but a thoroughly practical one, is its use in photography. This has been entirely developed within the past year or two; indeed the first kite-photograph taken on the American continent was one made by Mr. Eddy's camera on May 30, 1895. Although some attempts in this direction had been previously made in Europe, this was the first clearly focused kite-photograph obtained. The previous ones had been blurred, owing to defects in the devices for swinging the camera apparatus from the kite-cord, and for loosening the shutter. Mr. Eddy's apparatus will be better understood from the accompanying cut than from any description. In a general way it is a wooden frame capable of holding the camera, and terminating behind in a long stick or boom, by means of which the camera is made to point in any desired direction or at any angle. This is arranged before sending up the apparatus, the boom being properly placed and held in position by means of guy cords from the main kite-line. A separate line hangs from the spring of the camera shutter, with which is also connected a hollow ball of polished metal supported in such a way that it will drop from its position, five or six feet through the air, when the camera cord is pulled. The purpose of this ball is to allow the operator on the ground to be sure that the camera has responded to his pull and that the desired photograph has been taken. He is assured of this, having given the pull, on seeing the flash made by the polished ball in its fall.

All this being arranged, it is only necessary to send the camera up to any desired altitude and pull the camera cord, in order to get photographs of wide-stretching landscapes, extensive cities, like New York, and panoramas of every description. Such photographs could not but be of the greatest value to geologists, mountain climbers, surveyors, and explorers. And they must possess particular interest for students of geography and for mapmakers.

POSSIBLE USE OF KITES IN WAR.

It is obvious, too, that kite-photographs might be of great value in time of war, since a detailed view of an enemy's lines and fortifications might be thus obtained; while at sea a perfected kite-photographing apparatus might be of great value in recording the approach of an enemy's ships. Mr. Eddy regards it as perfectly possible to send up a tandem of kites from the deck of a man-of-war, with a circular camera, such as has already been devised, attached to the main line, and an apparatus for snapping all the shutters simultaneously; and photograph, not only the whole horizon as seen from the deck of a vessel, but, because of the greater elevation, many miles beyond. A battle-ship provided with this photographing device would enjoy as great an advantage as if it were able at will to stretch out its mainmast into a tower of observation a mile high.

It is true that some of the lenses in the circular camera, the ones facing the sun, might give imperfect pictures; but in whatever position the sun might be, at least one hundred and eighty degrees of the horizon would be clearly photographed. And by taking such observations in the early morning, and again in the middle of the afternoon, it would be possible to cover the whole circuit, and thus be aware of the approach of an enemy's ships long before they would have been visible to a telescope used on the deck. In siich a circular camera each lens would be numbered, and the position of each would be accurately determined with regard to the points of the compass by the use of guy-cords stretching from the main line to the framework of the apparatus. Thus, on looking at the number of a lens, the photographer would immediate-

The kite buoy in service.

ly know from which direction any vessel whose image was shown might be coming.

Nor is the use of the kite in war limited to the services it would render in photography; it might easily do more than that, and become a most efficient and novel engine of destruction. As has been shown, it is merely a question of carpenter work to send up a tandem of kites that will swing a heavy load high in the air. Suppose that load were dynamite, with an arrangement for dropping it over any desired spot. Mr. Eddy suggests that this might be effected by means of a slow match made by soaking a cotton string in saltpetre, which would be lighted on despatching the load of dynamite, and would burn at a regular rate, say one foot in five minutes, so that the length of the match could be timed to meet the necessities of the case. On burning to its end, the match would ignite a cord holding the dynamite in a pasteboard receptacle, one side of which would fall down like the front of a wall-pocket as soon as the restraining cord was burned through; and immediately the dynamite in the box would be launched toward its destination. Mr. Eddy has already carried out an experiment similar to this, in setting loose from high elevations tiny paper aëroplanes. With a little practice he found he could start the slow match with such precision as to cause the aëroplanes to burst out into flight at any desired altitude. This interesting and beautiful experiment was performed for the first time by Mr. Eddy on February 22, 1893, when he sent off from a height of one thousand feet forty aëroplanes, their forward edges weighted with pins for greater stability.

Assuming such an arrangement made for discharging a load of dynamite, Mr. Eddy calculates that, with a twenty-mile breeze, six eighteen-foot kites would lift fifty pounds of the explosive a quarter of a mile in the air and suspend it over a fort or beleaguered city half a mile distant. It would thus be perfectly possible, supposing the wind to be in the right direction, to bombard Staten Island with dynamite dropped from kites sent up from the Jersey shore. It is evident that, for purposes of bombardment, a tandem of kites possesses several advantages over the war balloon. Kites are much cheaper. Then it would be far more difficult to disable them than to disable a balloon, since they offer a smaller mark to the enemy's guns; and even if one or two were destroyed, the others would still suffice to carry the dynamite. Finally, the kites may be sent up without risk to the lives of those who directed them, which is not the case with the balloons.

Another interesting and important application of the modern kite has been conceived by Professor J. Woodbridge Davis, principal of the Woodbridge

Boys' School, in New York, who is one of the most famous kite-flyers in the world, in addition to being a distinguished scientist and mathematician. It was Professor Davis who invented the dirigible kite several years ago, three strings allowing the operator to steer the kite from right to left at will or to make it sink to earth. Having perfected this curious kite, which is of hexagon shape, is covered with oiled silk, is foldable, portable, and has a tail, Professor Davis turned his attention to his more recent and important discovery of the dirigible buoy, which bids fair to do much to lessen the dangers of shipwreck. For months past Professor Davis, assisted by Mr. Eddy, has been experimenting on the Kill von Kull with this buoy, and has obtained most encouraging results. There are two kinds, both being designed to be attached to kite lines and drawn over the water by the power of the kite. The simpler variety is merely a long wooden tube about three inches in diameter and shaped very much like a gun projectile, with a cone of tin dragging behind to give steadiness. It is for use only when the wind is blowing in exactly the direction in which it is designed to send a message or carry a rope. It will be observed that, in a large number of cases when ships are driven on rocks, the wind is blowing toward the shore, and in such cases a line of kites would readily carry one of these buoys ashore with the important words inside or the still more important rope following after.

Not satisfied, however, with this buoy, Professor Davis sought some means of making kites draw a load across the water in any direction desired, regardless of the way the wind might be blowing; and, after much thought and calculation, he hit upon what is now known as the Davis buoy, an object that has become familiar to dwellers at Bergen Point and Port Richmond, from the frequent experiments on the Kill that have been carried on during the past year. This form of buoy is much larger than the other, being three or four feet in length; and its essential feature is a deep iron keel that projects below out of the block of wood forming the body. It is evident that this keel will tend to keep the buoy headed in any given direction; and stability of position is further assured by the presence of guy-ropes attached to the main line of the kite. Each buoy is provided with three of these ropes, which, by being lengthened or shortened, may cause the buoy to form any desired angle with the kite-cord, and to keep it. Professor Davis has entirely succeeded in making the kites drag the buoy along the water in various directions in the very strongest gales—in fact, under precisely the conditions that would assist when the buoys would be needed for lifesaving service from wrecks. And he is positive that, with further experiment, he will be able, by moving along the shore until a tacking angle is reached, not only to send lines, food, or messages to a disa-bled vessel from the shore, but to bring back by the same kites and the same buoy other lines and messages from the people in distress.

Considering the important offices of which it has already been proved ca-pable, and the possibility which these suggest of many other practical applica-tions, it is clear that the kite is no longer to be regarded as simply a toy. And this, in turn, suggests anew the familiar truth that, after all, nothing in this world is of small consequence.

When Mountains Blow Their Heads Off

Marvelous Facts in the Action of Volcanoes. —Some Observations by Prof. John Milne.

By Cleveland Moffett

September 1898

IN 1878, when Professor John Milne, then occupying the chair of geology and mining at the University of Tokio, was journeying over Japan describing its active volcanoes, he came to innocent old Bandaisan, about a hundred miles north of the capital, and for some time was in doubt whether to include her in his list or not. As far as he could learn, there was not a better behaved mountain than she in the whole empire; she never smoked, she never shook, and there were no traditions of her having been in eruption even at the most distant period. She simply rose out of her lonely valley, and went on, century after century, holding up the sky and troubling no one. She rose to the height of about a mile, and was calm and grand.

But peasants in the valley told of hot springs coming out from the base that brought poor people thither in numbers for their healing virtues, and when the Professor saw these springs he knew that he must look further, for where there is hot water there may be steam, and when steam gets into the bowels of a mountain many things may happen not provided for by the word "extinct." So he pressed up the mountain's sides, beautiful with verdure, and underneath the mosses and trailing vines he came upon scoriaceous lava, which is another sign. Then he went right to the top, up the steepest slope, and found as fair a spread of vegetation as the eye could rest upon; and presently two deer came bounding from the undergrowth as if to show him that there was no danger. Nevertheless, he found a crater underneath, a genuine

View of Krakatoa during the earlier stage of the eruption.

From a photograph taken May 27, 1883, and published in the report of the Krakatoa committee of the Royal Society, 1888, entitled "The Eruption of Krakatoa and Subsequent Phenomena."

volcanic crater, and without more searching he classed Bandaisan among the active volcanoes of Japan.

Then see what Bandaisan did. On July 15, 1888, ten years later, with no warning and for no reason that anyone can find out who does not know the secrets under the earth, she blew her beautiful green head off, and sent sixteen hundred million cubic yards of rock and earth—that is Professor Sekya's estimate—to arrange themselves in the valley beneath as best they might. There is little use trying to think of sixteen hundred million cubic yards of rock and earth; it is better to do some figuring, and this shows:

> (1) That if the mass blown away by Bandaisan at this time had been in nicely hewn fragments each the size of an ordinary street car, there would have been a train of these long enough to go five times around the earth.

(2) That if these fragments had been blown into great shells as large as the largest ship afloat, with a displacement of, say, 15,000 tons each, they would, if floated end to end, have bridged the Pacific from San Francisco to Yokohama.

Within three days of this startling justification of his conclusions as to Bandaisan, Professor Milne was at the scene of the disaster, and was the first person to make thorough and accurate observations of what had taken place. It is to him that I am indebted for the facts about this eruption, and also for photographs taken on the spot by his friend, Professor W. K. Burton.

A FURIOUS RIVER OF MUD AND STONE.

Now, this is what had happened. A river of "moya" or agglomerate, not lava, but a mixture of mud and stone, had poured down the valley at the rate of forty-eight miles an hour, and in twenty minutes had spread itself to a depth of one hundred feet over a region from twelve to fifteen miles long and from five to seven miles wide. When a river of mud travels down a valley at this rate, nearly a mile in a minute, a river as deep as a church, it is needless to say that Death rides on the wave for a quick garnering. That valley would have taken in the greater part of New York City, which is long and narrow, and had New York City been there at this time, some two million mortals would have sent their last breaths bubbling up through mud. As it was, only 401 persons lost their lives, because only 401 persons were there to lose them. The same is true of houses and buildings: whatever was in the valley was destroyed; and for miles beyond, in all directions, villages were wrecked by the air-blast, trees were stripped bare as if by a forest fire, and crops standing in the fields were flattened on the ground like threads for a loom.

Near Bandaisan is Lake Inawashiro, and from this point Professor Milne and his party, on the morning after their arrival, set out for the ruins. They started at daybreak, and explored until after dark, walking over a waste of steaming, slippery debris. They slid down banks of mud, not knowing what they should find at the bottom nor how they could get out again; they climbed over boulders like small cathedrals; they viewed the rebellious mountain from many points, and saw that its head was indeed missing, only a jagged neck showing here and there when the steam lifted. And they saw with amazement how the face of things was changed: everything bare and brown where carpets of green had been; houses gone, people gone, the valley buried in mud, and here, where dry land was, a new lake forming. This lake was caused by the sudden damming up of a mountain stream, and was destined to go on growing for two whole years, so that to-day it rivals Inawashira, and has actually caused the peasants in its vicinity to abandon farming and devote themselves to fishing.

There was one phenomenon observed by these first explorers which gave rise to much controversy. They found the plain, beyond the mud-swept valley, covered with conical holes several feet in diameter, that looked like small volcanoes. And some insisted that there had been minor eruptions here at the time of the big one, but their reasonings were presently overthrown by the discovery that at the bottom of each one of these holes, buried six or eight feet under the ground, were boulders from Bandaisan which had embedded themselves thus in falling. When it is considered that these boulders were of considerable mass, some weighing four or five tons, and that they had been hurled eight or ten miles from the summit, the velocity with

Bandaisan still smoking and steaming after it has blown its head off.

The ragged line at the top what was the neck of the mountain. This and the next are from photographs by W. K. Burton

which they must have struck the earth is seen to have been enormous. Indeed, it is the opinion of Professor Milne that they fell from a height sufficient to give them the maximum velocity that may be attained by bodies falling through our atmosphere, a velocity equal to that of falling meteorites, for it must be understood that the increasing resistance of the air puts a definite limit upon such velocities.

TWO KINDS OF VOLCANIC ERUPTION.

In our talk about Bandaisan, I naturally asked Professor Milne what were the causes of such an appalling catastrophe as this, and in explaining these causes he pointed out that there are two kinds of eruptions to be noted in the history of a volcano, those that build it up very slowly, and those that destroy it very swiftly, as if nature amused herself by piling up these great masses through the ages simply to see how quickly she could tear them down.

The eruptions that build up mountains, I understood, are periodical wellings over of molten lava, comparatively harmless. The others are violent explosions, occurring irregularly and bringing widespread destruction. It is easy to see how each streamingover of the lava makes the mountain grow, just as an icicle grows or a stalactite; each fresh outgush hardens as it pours, and forms a fresh shell of lava for other shells to form on. And, finally, when a certain height is reached—one, two, three miles—we may suppose the impelling force beneath no longer equal to the task of lifting this great column, and the crater crusts over at the top, and so generations pass, and men with their short lives and shorter memories say that the volcano is dead.

On the steaming slopes of Bandaisan after the explosion.

At the bottom of the picture is shown part of the new lake formed by the damming up of a mountain stream.

But the fires are there at the core, so much latent energy ready to be stirred; and if something stirs them, it is like rousing a thunderbolt. The fact that the natural vent above is blocked with the coolings of centuries only makes the discharge the more terrible when it comes, just as hard rammed bullets make the powder more effective.

I asked what was the cause that usually determines one of these explosions and rouses the volcano's latent energy, and I learned that in most cases it is the very same cause that makes a boiler burst—the sudden and excessive generation of steam when the hot part of a volcano comes in contact with water. This contact may be due to various causes, as, for instance, the readjustment of strata or materials beneath, so that a lake or watercourse is turned into the crater. It may even be due to an irruption of the sea, as at Krakatoa in 1883.

"Then does molten lava never come out in one of these violent explosions?" I asked.

"Sometimes it does; sometimes it does not. It did in 1873, when Asama, another Japanese volcano, blew its head off, and the lava track may still be seen along the face of the mountain like a huge black serpent. But in cases like that the lava does not well out; it is driven out by the steam, just as rocks are driven out."

"And when no lava comes out, where does the mud river get the liquid to make it flow?"

"Partly from the steam, partly from water it absorbs from springs and streams in its course. The mud river from Asama, for instance, lapped up two ordinary rivers as it went, so that no sign of them appeared thereafter."

"Is it likely, Professor, that there are volcanoes in the world at present that have been quiet for a long time, but will one day or another blow their heads off?"

"It is almost certain that there are."

"Some in Europe?"

"Many in Europe."

"Some in the United States?"

"Undoubtedly."

"Some in England?"

"Very probably, although there is no telling when they will do it. England has at least a dozen basal wrecks of volcanoes, mostly in the western Highlands, regarded as extinct, but Bandaisan has shown us what 'extinct' volcanoes will do. An 'extinct' volcano is very much like an old rusty gun—it may be loaded."

THE GREAT EXPLOSION AT ASAMA.

Next we talked about the explosion of Asama, the great one of 1783, which Landgrelle, a distinguished authority, regards as one of the most frightful eruptions in the history of volcanoes. There is special reason for referring to this mountain since its ragged shoulders, from which the head was blown at this time, were the scene a few years since of an interesting and rather hazardous experiment attempted by Professor Milne and a party of friends. Asama rises to a height of over 8,000 feet, and in its great paroxysm it sent down, so the records say, a river of mud from five to ten miles broad, that overwhelmed forty-two villages. "In some places," continue the records, "the mud was so hot that it did not stop boiling for twenty-four days. . . . In the Tonezawa River immense masses of lava remained red hot even in the river itself. . . . In Kurogano a stone 120 by 264 feet, one among many, fell into a river and formed an island. Two rivers were sucked up into the mud torrent and their places taken by dry land, and the noise of the explosion was like a thousand thunders. The lakes were poisoned, and fish sickened, the rivers were full of dead dogs, deer, and monkeys, with hair singed from their bodies."

The crater of this volcano, as it stands today, measures a mile and a quarter in circumference, and never ceases to belch forth pungent, strangling vapors of hydrochloric acid and sulphurous anhydride, to breathe which is to die. The depth of the crater has been a subject of endless discussion among the foreign residents of Tokio, some putting it at 1,000 feet, others at 8,000, and it was to settle this controversy that the experiment just referred to was undertaken. A party set out one day, headed by Professor Milne and United States Minister Edwin Dun, with no less an object than to sound Asama's crater. They took with them elaborate chemical and physical appliances, a great quantity of rope, and a number

View into the smoking mouth of Asama.

It was across the chasm shown in the picture that Professor Milne's party stretched a rope tackle in their attempt to measure the depth of the crater.

View of the summit of Bandaisan after the explosion.

The great rock at the right is a specimen of many, an large as houses or churches,
which were tumbled for miles in all directions.

of coolies to haul it. When they reached the edge of the crater, keeping carefully to the windward of the vapors, they proceeded to execute an idea of Minister Dun for measuring the depth, an idea that had been adopted after much discussion. First, with extreme difficulty, a rope was stretched across the crater, a distance of about 500 yards. Then a pulley was run out on this fixed line with another rope that could be lowered straight down (a thick wire was tried first, but it kinked and broke), and at the end of the vertical rope was made fast what the explorers called their "chemical and physical laboratory," that is, special thermometers, bits of metal and other substances that would fuse at various temperatures, pieces of red and blue litmus paper, etc.

Finally, when all was ready, the coolies were told to lower away, and the rope began to go down in the very thick of the vapor clouds, while all waited expectantly. Everything went well until a depth of 735 feet was reached, and then the experiment came to an abrupt and disconcerting end by the burning up of thermometers, rope, and everything. And that is the only attempt that has ever been made to penetrate the mysteries of Asama's crater.

THE GREATEST EXPLOSION EVER KNOWN.

Coming now to the explosion of Krakatoa, let me note that although we have here what is admittedly the most formidable volcanic convulsion of modern times, perhaps the most formidable in our whole history, yet the place of its occurrence was quite insignificant. Krakatoa on those memorable days in 1883, the 26th and 27th of August, was a poor neglected little island in the Strait of Sunda, between Java and Sumatra. No one lived there, no ships touched there, and in the presence of forty-nine towering volcanic mountains on the neighboring island of Java, some of them 12,000 feet high and most of them in chronic disturbance, no scientist had ever paused to observe the peculiar situation of Krakatoa with its one humble peak, measuring scarcely 3,000 feet. Had he given much heed, he would have made some important discoveries, notably that this humble peak was not the real volcano at all, but only a tooth in the ragged jaw of its vast crater, a crater that was largely submerged, and included not only the island of Krakatoa, but several other islands in the Strait of Sunda. And he would have seen that here, at some time in the dim past, had stood a great mountain that may have joined Java and Sumatra, and that certainly had a girth of twenty-five miles at its base and a summit towering with the best of them. That was the real volcano Krakatoa, after the work of its building up with lava layers had been completed, and before the phase of its self-destruction had begun. Then, in the pride of her strength, Krakatoa proceeded to tear herself to pieces; she blew her head off, she blew her shoulders off, she scattered her body far and wide, and finally left herself only a "basal wreck," in the words of Darwin, to rest upon, and that half under water. All this the scientist would have discovered, and also that, broken and disfigured though she was, Krakatoa still stood at the intersection of two great lines of volcanic energy, and therefore marked the most dangerous volcanic focus on the surface of the earth.

But all this came as after-knowledge, and the giant force imprisoned in that unheeded crater was allowed to rend asunder its fetters with a quaking of the earth and a blazing of the heavens before any suspicion of its presence went abroad. For nearly 200

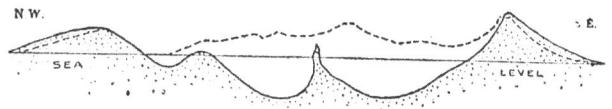

Outline of the crater of Krakatoa as it is at the present time.

The dotted line indicates the portions blown away in the paroxysmal outhurst of August, 1883, and the changes in form of the flanks of the mountain by the fall of ejected material upon them. Reproduced from "The Eruption of Krakatoa and Subsequent Phenomena."

years Krakatoa had done nothing; then on Sunday morning, May 20, 1883, she began to rouse herself, merely a matter of steam and falling ashes, with a roaring heard plainly in Batavia, a hundred miles away. Then followed three months of menacing prelude, as if she wished to give the world fair warning. Then, on the 24th of June, a second crater opened. Soon after this a third crater opened.

The low-lying walls of the craters had at last given way in many places, and there were white hot chasms below the level of the sea sending up to the waves their hissing challenge. Then thousands of tons of water surged downward, and the fight was on. This was Sunday afternoon, August 26th. For the first few hours the fires of the earth made short work of the sea, driving it back in splendid explosions that came every ten or twelve minutes. Each explosion sent up black columns, miles in height, steam and smoke and ash and pumice, all the scum and debris on the surface of the molten lake, and drove back the sea in great waves. Soon the darkness of night settled over Java and Sumatra and over vessels sailing in those waters, and through the darkness at intervals was seen the glory of Krakatoa, a terrifying glory. "From a distance of forty miles," says an eye-witness on a ship, "it looked like an immense wall, with bursts of forked lightning darting through it and blazing serpents playing over it." These bursts of brilliancy were the regular uncoverings of the angry fires.

As the hours passed, the sea gained an advantage through fresh breaks in the crater walls that offered new points of attack. The explosions became more and more frequent until about midnight they sounded to the people of Batavia and Buitengong like one continuous roar, the noise making it impossible for the inhabitants of these places to sleep. It was generally believed that a heavy cannonading was going on in the immediate vicinity, though why, no one could imagine. The concussion shattered stone walls, upset lamps, and tore gas meters from their fixings. And yet Batavia is as far from Krakatoa as London is from Portsmouth.

And all through that Sunday night electricity did wonderful things in the heavens, and sailors saw balls of fire resting on the mastheads of their ships and at the extremities of the yardarms, and in some cases lightning struck the mainmasts. The climax came the next morning at about ten o'clock. For some hours the explosions had been more violent, though at longer intervals; the sea had made the fire retreat, but the fire had checked the farther passage with walls and floors of hardened lava. When these blew up, it was like blowing up the eternal foundations. And the hardest shock was yet to come. Did

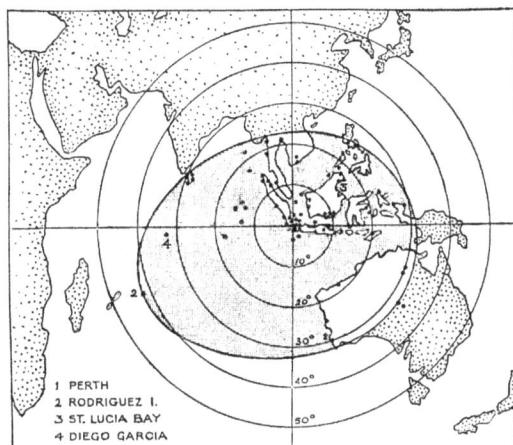

Map showing the places where the sounds of the great Krakatoa explosion were heard.

The oval indicates approximately the area over which the sounds were heard. The distance from Krakatoa to Perth is 1,902 miles; to Rodriguez Island, 2,908 miles; to St. Lucia Hay, 1,110 miles; and to Diego Garcia, 2,207 miles. From "The Eruption of Krakatoa and Subsequent Phenomena."

the earth open in one gigantic fissure and call the sea down for a final desperate encounter, or was there a sudden subsidence of strata to fill in the hollows left by what had been ejected? Not even the wisest scientist can say. But there came an explosion so loud, so violent, and with such far-reaching effects, that it made what had gone before seem as child's play in comparison, and made all other explosions known to the earth in historic times dwindle into insignificance.

To begin with, this explosion set in motion air waves that traveled around the earth four times one way and three times the other; that is, they disturbed every self-recording barometer on the globe no less than seven times. They traveled around the earth once in about thirty-six hours, or at the rate of 700 miles an hour, which is somewhat slower than sound waves travel. For it must not be supposed that these airwaves produced sound; their periods of vibration were too long for that; in other words, their sounds were too low for our range of hearing. Those that went in the direction of the earth's rotation, that is, from west to east, traveled about twenty-eight miles an hour faster than the waves which went in the opposite direction.

Besides these inaudible air waves, there were others of shorter vibration, that came within our range of hearing. These waves carried the sounds of the last terrible explosions over distances far beyond anything else known in human experience of sound transmission. All over Sumatra and Java the sounds were distinctly heard, which is as if all the people in New York should hear an explosion in Boston. That, however, is nothing. A resident at St. Lucia Bay, Borneo, 1,116 miles distant, writes: "The noise of the eruption was plainly heard all over Borneo."

This last was as if people in Chicago had been frightened by a noise in New York. But still this is nothing. From Tavoy, Burmah, 1,478 miles distant, they sent out the police launch in alarm; and Staff Commander Coghlan, R. N., writes from Perth, West Australia, 1,902 miles distant: "This coast has been visited (August 27th) by sounds like the firing of guns inland." And Mr. Skinner, of Alice Springs, South Australia, 2,233 miles distant, writes: "Two distinct reports similar to the discharge of a rifle were heard on the morning of the 27th, and similar sounds were heard at a sheep camp nine miles west of the station, and also at Undoolga, twenty-five miles east." At Diego Garcia, an island in the Indian Ocean, 2,267 miles distant, the people heard sounds from the east so distinctly that they thought it must be a ship in distress. And finally, Mr. James Wallis, chief of police in the Island of Rodriguez, which is almost across the Indian Ocean, 2,968 miles from Krakatoa, writes: "Several times during the night of the 26th-27th, reports were heard coming from the east, like the distant roar of heavy guns." This was as if a noise in Philadelphia had been heard in San Francisco.

Summing up the results of many reports like the above, it stands as certain that the Krakatoa explosion was heard over a sound zone covering one-thirteenth of the earth's entire surface; also, that the sounds, as is seen from the accompanying diagram, were carried much farther toward the west than toward the east, owing probably to the fact that a strong wind was blowing at the time.

Coming next to the sea waves sent from Krakatoa, the damage done by these was enormous. Two lighthouses in the Strait of Sunda were destroyed, all the towns and villages on the shores of Java and Sumatra bordering the strait were destroyed, all the boats and vessels on the same shores were destroyed, and 36,380 lives were lost. The tidal wave which started at ten o'clock was the one which wrought the worst destruction. Its average height when it struck the shores of Java and Sumatra is estimated at fifty feet, but in many places it is known to have been much higher than that. At Merak, on the Java coast, where there is a funnel-shaped bay that may have heaped the water up, the wave is said to have reached a height of 135 feet. And a man-of-war, the "Berouw," lying off the Sumatra shore, was carried a mile and three-quarters inland up a valley and left in a forest thirty feet above sea level.

These sea waves traveled across the Indian Ocean in all directions, and were recorded by tide gauges at Colombo, Ceylon, 1,760 miles distant; at Bombay, 2,700 miles distant; and at Cape Horn, about 5,000 miles distant. That is, they washed the southern coasts of Asia and the eastern coasts of Africa. Their average rate of transmission was about 350 miles an hour; their average height, as shown by the gauges, was from six to eighteen inches.

Coming now to other effects of this great explosion, it is established on the evidence of many officers of ships and dwellers on islands, that on this day a large part of the Indian Ocean was showered with lava dust and lava mud to a depth of several inches. This applies to an area of, perhaps, about half a million square miles, but in the immediate vicinity of Krakatoa, say within a hundred miles, the sea was so thick with fallen lava dust and debris that vessels pushed through it with great difficulty, as if they were passing through a field of broken ice. As for the Strait of Sunda, it was rendered quite impassable with mud and pumice, which is as if Channel steamers were blocked on their way to the Continent, because the Straits of Dover were covered with mud a foot or more deep. In a word, the mass of mud and ashes and lava dust blown out of Krakatoa into the air would have formed a solid cube a mile and a quarter in each dimension. That is four or five times more than Bandaisan threw out.

A great quantity of the finer dust projected into the air remained in suspension there for over a year, and by a refraction of light caused the red and purple sunsets, the blue moons, and the copper suns that were seen all over the world from September, 1883, to the close of 1884, and that caused so much discussion and alarm. The whole northern portion of the island, much the greater portion, with an area of nearly six square miles and an average height above sea level of 700 feet, was submerged, and remains so to this day, under 150 fathoms of water. Two new islands had thrust up their heads, and the whole configuration of the channel was altered. All of which confirms one in the opinion that, when this old earth begins to fire off her heavy artillery—that is, blow the heads off her mountains—it makes human battles and explosions in powder and dynamite factories and the like look rather small.

Possibility of Life on Other Worlds

Recent Discoveries Bear Out Old Arguments That Other Planets May Be Inhabited.

By Sir Robert Ball, Professor of Astronomy at the
University of Cambridge, England

July 1895

NOTWITHSTANDING the wonderful advances in scientific methods which have been effected in recent years, a great problem still remains unsolved. We are still as far as ever from having attained any definite answer to the question as to whether life can exist on any of the other worlds. Vast as has been the progress in knowledge since the days when Whewell and Brewster discussed the question of possible inhabitants in other planets, a writer in the present day finds the problem which they attempted still hopelessly beyond his reach in so far as any determinate conclusions are concerned.

But it seems worth while to take up the question afresh, inasmuch as some of the old arguments have acquired increased significance in consequence of later discoveries, while others are now seen to have lost something from the same cause. I propose, accordingly, to set forth some account of the present state of the argument, and to note whatever additional importance it may have acquired since the days when the habitability of other worlds was discussed by Brewster.

The standard argument in support of the belief that certain other planets might be inhabited was of this kind. It was noticed that the sun lies at the centre of a system of bodies which revolve around it, and that among these bodies the earth holds an intermediate place. It is nearer to the central luminary than are some of the other planets, while, on the other hand, it is more remote than others. The warmth and light received by the earth from the sun would, therefore, be greater than that received by some planets and less than that received by others. If some of the planets are much larger than the earth, then it must be remembered that other members of the same system are smaller than our globe, and that some of them are very much smaller. It was also pointed out that the earth in another respect is, as it were, a fair average specimen of a planet. Some of these bodies have moons revolving around them. It is quite true that Jupiter, Saturn, and Uranus are more richly endowed with attendant globes than is the earth; but then Mercury and Venus appear to be unprovided with any moons. It was thus seen that in the matter of satellites, as well as in dimensions and in situation, our globe is an intermediate one in the system. This conclusion was confirmed by the subsequent discovery that Mars had a pair of satellites, and Neptune a single one. Indeed, the claims of the earth to be a typical planet might be pushed still farther. A notable characteristic of a planetary globe is its density, that is to say, its weight in comparison with the weight of a globe of water of equal dimensions. Here, again, our earth appears in the light of a fairly representative object. It is much lighter, no doubt, bulk for bulk, than some of the other planets. It is, on the other hand, much heavier than others.

It is also noticeable, in this connection, that our globe is surrounded with a copious atmosphere, and this is an attribute which, of course, stands in an obvious and specially important relation to the question of the earth as an abode of life. Those who pondered on the possibility of life on other worlds could not fail to be struck by the fact that some of those other worlds were also surrounded by atmospheres. If these atmospheres, in certain cases, were excessively dense and abundant, and in others greatly attenuated, this circumstance alone would tend once again to illustrate the intermediate rank, so to speak, of our earth as a member of the planetary system.

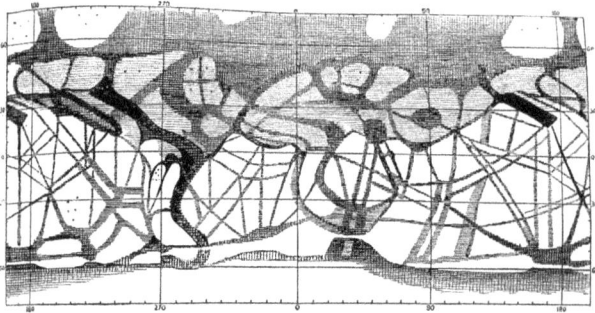

Bianchini's map of Venus.

From Webb's "Celestial Objects for Common Telescopes."

The argument then ran in this wise. Regarding our earth as a globe which constitutes a member of the solar system, it can hardly be said to possess very extreme attributes. It does not appear to be marked out in any specially distinctive manner which would qualify it rather than certain of the other globes for becoming suitable abodes for life. The qualities which the earth possesses are, generally speaking, conferred upon it in degrees intermediate to those in which other globes of the system are endowed with similar qualities. As the earth was inhabited, it would seem only reasonable to assume that in this respect also it was not exceptional, and that in all probability the other globes, some of them, or many of them, were also fitted for the abode of life, suitably adapted to the conditions which each globe had to offer.

Such was, in outline, the famous argument which was presented half a century ago in support of the conclusion that in all probability certain other planets besides our earth contained organic life. It is worth while to see how far the present state of our knowledge affects the validity of this argument. That it does so, cannot be

Sir Robert Ball, LL.D., F.R.S.

From a photograph by Elliott & Fry, Baker St., London.

questioned. I believe, on the whole, the argument has been strengthened by modern research, though it must be admitted that in some respects its efficiency has been impaired.

CORRESPONDENCE BETWEEN THE EARTH AND OTHER PLANETS IN MATERIAL.

We can, indeed, in these present days, bring forward a striking point of relationship between the earth and the other planets as to which the earlier writers had no information. Had they been aware of it they would certainly have regarded it as greatly strengthening the contention that it was reasonable to presume that the planets must be inhabited. But in those days philosophers had little notion that so astonishing a fact would ever be demonstrated as that the material constituents of the earth were in a great measure identical with the materials constituting the sun. They did not know that the ele-

Signor Schiaparelli's chart of Mars, showing the markings called by him the "Double Canals."

From "Old and New Astronomy," by Richard A. Proctor. Longmans, Green & Co., publishers, London and New York.

mentary bodies in the earth were substantially the same as the elementary bodies which make up the mass of the great luminary. It is, no doubt, quite true that we are not as yet able to affirm, with any absolute certainty, that the materials from which the planets, such as Venus or Mars, have been built, are actually the same kind of materials as those which make up the earth. Our knowledge, indeed, stops short of this point. We can pronounce on the substantial identity of the solar materials with the terrestrial materials, because in the former case the bodies are so greatly heated that they are in the gaseous state. Spectroscopic methods are therefore available for determining their identity with the glowing vapors of the same substances as we have them on the earth. But the planets are not incandescent. Our spectroscopes may, indeed, to some extent, inform us as to the constituents of the planetary atmospheres, but the actual solid portions of the planets cannot be analyzed by any means at our disposal. There is, however, no reason to think that the elements of which the planets are composed differ considerably from the elements of which the earth is made. For most astronomers now admit that the sun and the planets have had a common origin from some primitive nebula, and as we verify this theory by showing that the earth and the sun are substantially of the same constituents, it seems impossible to doubt that the substances which form the earth are largely, if not wholly, the same as the substances out of which the planetary globes have been fashioned. A striking confirmation of this doctrine of material uniformity is presented by certain of the comets which belong to the solar system. It is quite true that such objects have, so far as physical condition goes, no resemblance to planets. It is, however, sufficiently remarkable that comets appear to be composed of materials resembling those of which our earth has been made. For these bodies happen to be, in part at least, of such a gaseous nature that we are enabled to submit them to spectroscopic analysis. They have thus been proved to contain some of the most important terrestrial elements.

It is therefore plain that the ancient argument in support of the notion that some of the planets might be tenanted with life can be considerably reenforced by modern discoveries. For it may now be regarded as practically certain that various elements known on this earth are present in the planetary bodies. We thus see that the components necessary for the physical framework of living creatures may, in all probability, be as abundantly provided upon some of the other planets as they are on the earth.

DISTRIBUTION AMONG THE PLANETS ELEMENTS IMPORTANT TO LIFE.

In this connection it is instructive to bear in mind what is known as to the distribution of those particular elements in space which appear to be most characteristically associated with the manifestation of life. No result of spectroscopic research among the heavenly bodies has been more remarkable than that which demonstrates the extraordinary abundance with which the element hydrogen is diffused throughout the universe. It is, of course, one of the commonest elements of the earth, entering, as it does, into the composition of every drop of water. Hydrogen is also a constituent part of a vast number of solid bodies; but the remarkable circumstance for our present purpose is that this same element is found in profusion elsewhere. Surrounding that visual glowing globe of the sun there is an invisible atmosphere, of which hydrogen is one of the most prominent components. A like conclusion is drawn from the spectra of many of the stars. In the case of certain specially white and brilliant gems, of which Sirius and Vega may be taken as the types, the chief spectroscopic feature is the extraordinary abundance in which hydrogen is present. Even in the dim and distant nebula; gaseous hydrogen is the constituent more easily recognized than any other which they may possess. Indeed, it may be affirmed that we do not know any other substance which is so widely diffused as hydrogen. It need hardly be said that this gas is an important constituent in those compound bodies with which life is associated. In that somewhat grewsome exhibition, which shows the actual quantities of the several elements of which an average human body is composed, the bulk of the hydrogen forms one of the most striking items; and indeed in connection with all forms of animal and vegetable life, hydrogen is of primary importance. In the argument from analogy for the existence of life in other worlds, it is significant to note that an element associated in such an emphatic manner with the manifestation of life here should now be shown to be widespread through the universe.

In like manner carbon, which is, of course, an essential factor, in organic substances, has been demonstrated to exist in other parts of the solar system. The most striking illustration of this fact is presented in the case of the glowing solar clouds, which there is now good reason to believe are due to carbon. Many of the comets exhibit lines in their spectra characteristic of the same element. If these bodies, as has been often supposed, are drawn by solar attraction from the remotest parts of space, the carbon which they bear testifies that this element is present through a wide extent of the universe. Here, again, modern research has gone far to strengthen the argument as to the possible existence of life elsewhere. It has shown the cosmical nature of that particular element which, if not itself the veritable abode of life, seems to be, at all events, a constituent thereof.

Illustrations of the material identity of the several globes in space might be extended. Have we not been told that a diet absolutely devoid of salt would be fatal? Now the salt, or, at all events, the sodium which forms its characteristic part, is not merely confined to the earth. The famous D line in the solar spectrum tells us that the same element abounds in the sun. Nor is this important element confined to the solar system. We have ample testimony as to the wide diffusion of sodium in stellar depths.

The iron which enters so largely into the framework of things material enters, as is well known, in no inappreciable quantity into the structure of the human body. Is there not some story of the materials for a medal of pure iron having been extracted from the mortal remains of some illustrious individual? At all events, iron in many ways, or in various combinations, is often associated with organic phenomena on the earth. It is, therefore, material to observe that this element, like others which I have mentioned, appears to be very widely distributed through space. It has been proved that many hundreds of lines in the solar spectrum must be attributed to the presence of an abundant iron atmosphere surrounding the heated solar globe. Even such distant stars as Aldebaran or Arcturus have been made to disclose the fact that iron enters into their composition in a very significant manner. If, therefore, there should not be life in the other planets, its nonexistence cannot apparently be attributed to the absence of such suitable materials as life requires to build up its physical abode. So far as our knowledge goes, we feel constrained to admit that such materials are certainly present on other globes besides the earth.

At the same time, it is right to call attention to the fact that we are obliged to use great caution in any conclusion we may draw as to the space distribution of another element of much significance in the vital phenomena of this earth. I allude, of course, to oxygen. I do not, indeed, say that there can be any good reason to doubt that oxygen does really exist in other celestial bodies. In all probability the life-giving gas is just as abundant on many other globes as we find it to be on this one. At the same time, it is proper to remember that a widely extended distribution of oxygen has not been demonstrated in the same emphatic manner as has the existence of the other elements to which I have referred. The dearth of reliable testimony as to the cosmical distribution of oxygen may be attributed not so much to the actual absence of that element from other bodies, as to the unsuitability of the means at our disposal for detecting its presence upon them. I need not go farther into this point than to remark that certain well-marked lines in the solar spectrum had been attributed to oxygen, and they were no doubt correctly so attributed. It was, however, proved by Janssen that the oxygen which caused these lines, or a great part of them, did not exist in the sun, but that the lines were largely, if not wholly, due to the oxygen in the earth's atmosphere. This is not to be taken as a proof that there is no oxygen in the sun. It merely says that its presence there has not been as yet conclusively demonstrated.

This weakness in one link of the chain of evidence does not, however, seriously impair the general conclusion already mentioned, that the substratum of material necessary for life exists on other globes besides the earth. I will only add that the element calcium, which is of essential importance in the shells or the coral of the lower animals, or in the skeletons of the higher, is also one of the elements widely distributed through space.

DIFFERENCES BETWEEN THE PLANETS IN THE MATTER OF TEMPERATURE.

We have thus seen that in one important respect the progress of modern research has strengthened the ancient argument from analogy in support of the belief that there is life on other worlds besides this one. It is right now to mention how, in another way, modern investigation has tended to impair that argument, or, rather, I should say, to limit its application. Various lines of reasoning have rendered it almost certain that in the matter of temperature the several planets present considerable varieties and contrasts. I do not here refer to the temperature of the surface of the planet, which is the result of sun-beams which fall upon it. No doubt there are individual peculiarities of each planet from this cause, the effect of which will be presently referred to. But what I am now discussing is rather the internal heat of the several globes of the system. It seems to be

Jupiter, as seen through a powerful telescope.

From "Old and New Astronomy," by Richard A. Proctor. Longmans, Green & Co., publishers, London and New York.

generally true that the larger the dimensions of a planet the greater isthe internal heat which it still possesses. Into the reasons of this we need not now enter; suffice it to remark, that the great globe of Jupiter in this respect offers a very marked contrast to the earth. It seems to be highly probable, if, indeed, it be not certain,that Jupiter is at the present time heated to a temperature, at its surface, greatly in excess of the temperature of the surface of the earth. We cannot, indeed, assign an actual value to the temperature of Jupiter, but there seems little doubt that it must be so great as to preclude the possibility of that globe being the abode of any types of life like those which flourish on the earth. It is, no doubt, just conceivable that living beings of some strange and unknown fashion might endure the conditions which Jupiter appears to present; but I do not know anything which would make such a view likely. What we have said about Jupiter may, with certain modifications, apply also to Saturn, and in some degree to Uranus and to Neptune. It seems impossible that any of these great planets are at present abodes of life in any sense which is comprehensible to us.

There is reason to think that, so far as internal heat is concerned, the planet Mars, as well as Venus and Mercury, occupies much the same position as the earth. In all four cases the internal heat may be said to be non-existent, in so far as its present effect on any manifestations of life are concerned. The superficial temperatures which these globes present, and the climates that they enjoy, must be attributed primarily to the heat received from the sun; of course, the actual effect on each globe is profoundly modified by its atmosphere, as well as by its distribution of land and water.

The four globes just named are at such varied distances from the sun that the amount of heat which they obtain will differ considerably. Mars can only get a smaller allowance of sunbeams than the earth, while Venus will receive mere, and Mercury a good deal more. If we represent the average intensity of sun heat as it arrives at the earth by 100, we shall find that the intensity at Mars is no more than 43. Venus receives a share which may be represented by 191, while Mercury would get as much as 667. At the first glance it might be thought that these figures must necessarily imply vast climatic differences between the different globes. I am certainly not going to deny that this is so. Indeed, it seems to be extremely probable that there may be astonishing differences between the climatic circumstances of the planets. But what I want to insist upon at this moment is that the condition of a planet as to climate is not merely a matter of sunbeams. A very important element consists in the extent of the atmosphere with which that planet is invested. There can be no doubt as to the presence of an atmosphere around Mars, and of another around Venus; but we have no reason to think that these atmospheres, either in density or in composition, resemble that which envelopes our earth. The atmosphere around Mars, indeed, appears to be far less copious than that with which our earth is provided. This much, at least, we conclude from the translucency of the environment which permits us to study the details of Mars with far greater clearness than a Martian astronomer who was trying to survey our globe would be able to obtain through the comparatively dense medium interposed by our skies.

THE EFFECT OF ATMOSPHERE ON THE CLIMATE OF A PLANET.

The character of the atmosphere of a planet will exert a marked effect upon the temperature and the climate of its globe. The abundance of that atmosphere and the proportion in which it contains watery vapor, or possibly other vapors, will all tend to modify the degree in which sun heat is admitted, and the degree in which, when admitted, it is retained. It would be quite possible for two globes enjoying equal shares of sun heat to have, nevertheless, totally unlike temperatures and climates in consequence of atmospheric differences. We know, also, that the distribution of land and water has a marked effect upon climate. It was the contention of Lyell, in his famous book, that the changes of climate in the course of geological time were mainly due to altera-tions in the relative positions of land and water. The mention of this will, at least, remind us that climate depends upon other elements besides sun heat and atmosphere.

The significance of these considerations in connection with our present subject can hardly be overestimated. A globe may at first sight appear to be too far from the sun to enjoy sufficient light and heat to make life endurable or possible. It may nevertheless happen that, by some suitably contrived atmos-phere and some special configuration of land and water, such a globe may pos-sess regions endowed with a mild or even a genial climate. On the other hand, a globe which was placed so close to the great source of light and heat that its inhabitants, if unprotected, would be submitted to an unendurable scorching, may yet be fitted with an atmosphere which shall render it sufficiently adapted for life, notwithstanding its apparently unpromising circumstances.

In illustration of the important climatic effects of an atmosphere, I need do little more than cite the case of the moon. Our satellite is practically at the same distance from the sun as is the earth, and in its case, also, internal heat has no present effect on the temperature of its superficial portions. It would, therefore, seem that so far as sun heat is concerned, the moon must be in much the same condition as the earth. But if we thence deduced the inference that the temperature conditions prevailing on our satellite bore any resemblance to the temperature conditions prevailing on the earth, we should make a great mistake. Observations of the moon's heat show that its surface is exposed to a tremendous range of temperature, extending to hundreds of degrees. It has been demonstrated that the temperature of the moon under the full glare of the sun rises to a point in excess of that of boiling water, while it is equally cer-tain that when the sunbeams are withdrawn the temperature of the moon sinks to a point far below that with which any Arctic explorer has made us ac-quainted. Here, then, is a globe fed just as we are with sunbeams, and yet un-dergoing tremendous vicissitudes of climate entirely surpassing any changes endured by the earth. The climatic difference between these two neighboring globes is certainly connected with the fact that the moon has very little atmos-phere, even if it be not completely destitute thereof. Our atmosphere acts as a climatic regulator. It reduces the degree in which the intense fervor of the sun affects the earth, and it mitigates the rigor of the cold to which the earth would be exposed when the sunbeams are withdrawn. Such an ameliorating agent is absent from the moon, and hence arise those violent extremes of its climatic condition. We thus see what potent factors the existence and the extent of an

Saturn and his rings.

From "Old and New Astronomy," by Richard A. Proctor. Longmans, Green & Co., publishers, London and New York.

atmosphere become in determining the nature of the climate that a planet is to have. We do not know enough regarding the atmospheres of Mars, Venus, and Mercury to be able to draw any certain conclusions with regard to their climates. But this much we may at least affirm, that it seems quite possible for the different influences we have named to go a long way toward neutralizing the contrasts which the climates of these globes would otherwise present in consequence of the different supplies of sunbeams that they receive at their actual solar distances. So far as mere climate is concerned, it seems quite possible that appropriate atmospheres and land distributions might be adjusted on the earth and Mars, Mercury, and Venus, in such a manner that certain organic types might be common to all the four globes.

Of course the presence or absence of water on a potential world must be a very material element in deciding as to whether life can exist thereon. The absence of water from the moon, for instance, must be at once admitted to be incompatible with the existence of life on that globe, in so far, at least, as the word life conveys to us any intelligible meaning. But though there is no water to be discerned at present on our satellite, yet it would seem highly probable that other globes may not be similarly destitute. One of the most striking features which Mars presents when that planet is placed in a favorable opposition, consists in his wonderful Arctic region of white material. This seems to grow as the winter advances on Mars, and decreases when summer reigns on that hemisphere of the planet which is exposed to us. Now we should certainly be going beyond the actual extent of our knowledge were we to affirm that what we see on Mars is certainly ice or snow, similar to that which we find in our own Arctic regions. It seems, however, hardly possible for us to frame any other supposition which could be reconciled with the facts. Indeed, the whole appearance of the planet makes it highly probable that water is quite as important a factor in the constitution of that globe as it is in our own.

Venus is so circumstanced in regard to the position which it has relatively to the earth that we are not able to examine it with the same degree of success as that which attends the study of our neighboring planet on the other side. It would appear, however, from the observations of Trouvelot, that the poles of this planet are also characterized by caps of white material, which remind us of the polar condition of our own earth, as well as of Mars. We do not see Mercury sufficiently well to form any conclusion as to whether it may possess similar features. The clouds of Jupiter, doubtless, also contain water, even if they are not entirely composed thereof; though, for the reasons already assigned, it seems quite unlikely that there can be any life on that globe.

RELATION BETWEEN THE SIZE OF A PLANET AND THE POSSIBILITY OF LIFE.

In the absence of any definite knowledge as to the composition of the atmospheres by which the planets are surrounded, or as to the climates which they enjoy, it would certainly be idle for us to speculate as to how far they might possibly be tenanted by creatures resembling those found on this earth. It would also be impossible for us to form any conception as to the biological characteristics of creatures which would be adapted for residence on the several planets. There is, however, one merely mechanical matter which may be usefully mentioned, inasmuch as it depends on considerations which admit of demonstration. We are able to weigh the several planets. Indeed, the problem is a comparatively easy one when applied to those bodies which are attended by satellites, inasmuch as the movements of the satellites contain indications of the weights of their primaries. But even when a planet has no satellites, it is still possible for an astronomer to find the weight of a body by the effect which its attraction produces on other planets. But the weight of a planet must stand in important relation to the framework of the organisms which are adapted to dwell upon it. Let me try to make this clear by a few illustrations.

Suppose that a planet, while still retaining the same size, was to be greatly increased as to its mass. The consequences would be felt very seriously by all organized creatures. The most immediate effect would be to increase the apparent weight of everything. If, for instance, a globe the same size as the earth possessed double the mass of the earth, the effect would be that the weight of each animal on the heavier globe would be double that on the earth. A horse placed on the heavy globe would be subjected to a load which would oppress him as greatly as if, while standing on our earth as at present constituted, he bore a weight of lead on his back which amounted to as many pounds as the animal itself. Each leg of an elephant would be called upon to sustain just double the not inconsiderable thrust which at present such a pillar has to bear. A bird which soars here with ease and grace, would find that the difficulty of such movements was greatly increased, even if they were not wholly impossible, on a globe of equal size to the earth, but double weight. It would seem as if flying animals must be the denizens of light globes rather than of heavy ones.

It is also easy to show that in general, other things being equal, the size of an animal should tend to vary in an inverse direction to that of the mass of the globe on which it dwells. At first it might be supposed that big animals might be most appropriately located on big worlds, and small animals on small worlds. No doubt there are so many circumstances to be considered, of which we are in almost complete ignorance, that any statements of this kind must be received with considerable caution. We may, however, assert with some confidence that, so far as our knowledge goes, the truth lies the other way. It is the small animals which are adapted for the larger worlds; it is the big animals which are adapted for the smaller worlds. The proof of this involves an interesting point.

LARGE WORLDS FOR SMALL ANIMALS, SMALL WORLDS FOR LARGE ANIMALS.

The argument is as follows: Suppose that an animal on this earth, as it is at present, were to have every dimension doubled. To take a particular instance, conceive the existence of a giant horse which was twice as high and twice as long, in every feature and detail, as an ordinary horse. It is obvious that, as all three dimensions of the animal are doubled, its volume, and therefore its weight, would be increased eightfold, and the weight that would have to be transmitted down each of the four legs would be increased eightfold.

Each leg of the giant horse would, therefore, have to possess eight times the weight-sustaining power that would suffice for the leg of the ordinary horse. As the proportions are supposed to have been observed throughout, the leg of the giant horse would be, of course, considerably stronger than that of the ordinary horse, but it would not be so much stronger as to enable it to accomplish the task it would be called on to perform. The section of the leg of the giant horse would, no doubt, be double in diameter that of the normal individual. This would imply that the area of the section was increased fourfold. But we have seen that the weight transmitted was increased eightfold. Study the effect of this on the horse's hoof in contact with the ground. In the giant horse the area of the surface of contact would be four times as great as in the normal horse. As, however, the weight transmitted is eight times as great, it follows that this wear and tear on each square inch of the foot, and this is the proper way to estimate it, would be just twice as destructive in the giant horse as it would be in the ordinary animal. If, then, as we may well suppose, the foot of the latter is just adapted for the work which it has to do, then the foot of the giant horse would be incapable of withstanding the wear and tear to which it would be subjected. It follows that an effective animal, on the scale we have suggested, would be an impossibility on our earth; at all events, when the materials from which it was made were the same as those out of which our animals are fashioned.

Suppose this giant horse, instead of being left on this earth, were transferred to another globe, which only exerted half the gravitating effect experienced on the earth's surface, then the effort the animal would have to make in supporting its own weight would only be half that which it has to put forth here. The consequence is that the framework of the giant horse would, in such a case, have to support a weight which was not more than four times that of an ordinary horse standing on the earth. As the area of the bases of support in the large animal was fourfold that in the normal horse, it would follow that, area for area, there would be a pressure transmitted through the foot of the giant horse on the less ponderous globe precisely equal to that of the normal horse on the earth. The materials of which the big horse is built ought, therefore, to be able to sustain him effectively when he was placed on the light globe. It therefore appears that, so far as gravitation is concerned, the big horse would be better adapted for the light globe, and the small horse for the heavy one. More generally we may assert that, regarding only the point of view at present before us, the limbs of smaller animals would be better adapted for vigorous movement on great planets than would those of large creatures.

It is, however, proper to bear in mind the point to which attention was, so far as I know, first called by Mr. Herbert Spencer. He has shown that there are excellent biological reasons, quite independent of those mechanical considerations to which I have referred, why it would be impossible for an efficient animal to be constructed by simply doubling every dimension of an existing animal. The support of the creature's life has to be effected by the absorption of nourishment through various surfaces in the body. But if all the dimensions are doubled, the bodily volume, as we have already mentioned, is increased eightfold, and therefore its sustenance would, generally speaking, require eight times the supply that sufficed for the original animal. On the other hand, supposing the same scale to be observed throughout the animal's body, the available surface area for absorption of nourishment has only increased fourfold, and therefore each square inch would have to do double duty in the large animal. If, however, the surfaces are at present at full work, it would seem impossible that they should efficiently undertake double the work they now get through. On this account, therefore, a live animal would seem impossible on a simple specification of dimensions twice those of any existing animal. Great structural modifications of pattern would have to accompany the enlargement of bulk. This, be it observed, is wholly independent of all questions as to gravitation.

No reasonable person will, I think, doubt that the tendency of modern research has been in favor of the supposition that there may be life on some of the other globes. But the character of each organism has to be fitted so exactly to its environment that it seems in the highest degree unlikely that any organism we know here could live on any other globe elsewhere. We cannot conjecture what the organism must be which would be adapted for a residence in Venus or Mars, nor does any line of research at present known to us hold out the hope of more definite knowledge.

Where Man Got His Ears

By Henry Drummond, LL.D., F.R.S.E., F.G.S.

June 1893

ONE of the most humorous sights in nature, less common in America than Europe, is a snail wandering about with a shell on its back. The progenitors of snails once lived in the sea, and when they evolved themselves ashore they carried this relic of the water with them,—an anomaly which, seen to-day, seems as ridiculous as if one were to meet an Indian in Paris with his canoe on his back. But there are more animals besides snails that once lived in the water. If embryology is any guide to the past, nothing is more certain than that the ancient progenitors of Man once lived an aquatic life. As the traveller, wandering in foreign lands, brings back all manner of curios to remind him where he has been—clubs and spears, clothes and pottery, which represent the ways of life of those whom he has met, so the body of Man, returning from its long journey through the animal kingdom, emerges laden with the spoils of its watery pilgrimage. These relics are not mere curiosities; they are as real as the clubs and spears, the clothes and pottery. Like them, they were once a part of life's vicissitude; they represent organs which have been outgrown; old forms of apparatus long since exchanged for better, yet somehow not yet destroyed by the hand of time. The physical body of Man, so great is the number of these relics, is an old curiosity-shop, a museum of obsolete anatomies, discarded tools, outgrown and aborted organs. All other animals also contain among their useful organs a proportion which are long past their work; and so significant are these rudiments of a former state of things, that anatomists have often expressed their willingness to stake the theory of Evolution upon their presence alone.

Prominent among these vestigial structures, as they are called, are those which smack of the sea. At one time there was nothing else in the world but water-life; all the land animals are late inventions. One reason why animals began in the water is that it is easier to live in the water—anatomically and

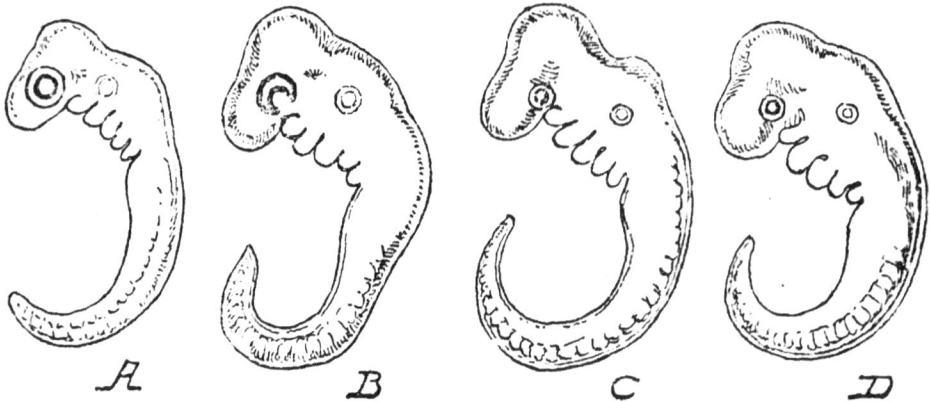

A. FISH.   B. CHICK.   C. CALF.   D. MAN.

Embryos showing gill-slits.

—from Haeckel's "Evolution of Man."

physiologically cheaper—than to live on the land. The denser element sup-ports the body better, demanding a less supply of muscle and bone; and the perpetual motion of the sea brings the food to the animal, making it unneces-sary for the animal to move to the food. This and other correlated circum-stances call for far less mechanism in the body, and, as a matter of fact, all the simplest forms of life at the present day are inhabitants of the water.

A successful attempt at coming ashore may be seen in the common worm. The worm is still so unacclimatized to land life that instead of living on the earth like other creatures, it lives *in* it, as if it were a thicker water, and al-ways where there is enough moisture to keep up the traditions of its past. Probably it took to the shore originally by exchanging, first the water for the ooze at the bottom, then by wriggling among muddy flats when the tide was out, and finally, as the struggle for life grew keen, it pushed further and fur-ther inland, continuing its migration so long as dampness was to be found. Its cousin the snail, again, goes even further, for it not only carries its shell ashore but when it cannot get moisture, actually manufactures it.

When Man left the water, however,—or what was to develop into Man—he took very much more ashore with him than a shell. Instead of crawling ashore at the worm stage, he remained in the water until he evolved into something like a fish; so that when, after an amphibian interlude, he finally left it, many "ancient and fish-like" characters remained in his body to tell the tale. Now, it is among these piscine characteristics that we find the clue to

172

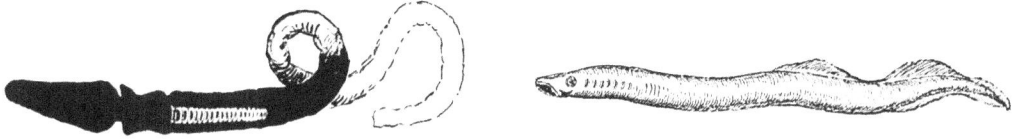

"Balanoglossus" (after Agassiz), and large sea lamprey (after Cuvier and Haeckel), showing gill-slits.

—from "Darwin and After Darwin," by Romanes

where Man got his ears. The chief characteristic of a fish is its apparatus for breathing the air dissolved in the water. This consists of gills supported on strong arches, the branchial arches, which in the Elasmobranch fishes are from five to seven in number and uncovered with any operculum, or lid. Communicating with these arches, in order to allow the water which has been taken in at the mouth to pass out at the gills, an equal number of slits or openings are provided in the neck. Without these holes in their neck all fishes would instantly perish, and we may be sure Nature took exceptional care in perfecting this particular piece of the mechanism. Now it is one of the most extraordinary facts in natural history that these slits in the fish's neck are still represented in the neck of Man. Almost the most prominent feature, indeed, after the head, in every mammalian embryo, are the four clefts or furrows of the old gill-slits.[1] They are still known in embryology by no other name—gill-slits—and so persistent are these characters that children have been known to be born with them not only externally visible—which is a common occurrence—but open, through and through, so that fluids taken in at the mouth could pass through them and trickle out at the neck. This fact was so astounding as to be for a long time denied. It was thought that when this happened, the orifice must have been accidentally made by the probe of the surgeon. But Dr. Sutton has recently met with actual cases where this has occurred. "I have seen milk," he says, "issue from such fistulae in individuals who have never been submitted to sounding." [2]

In the common case of children born with these vestiges, the old gill-slits are represented by small openings in the skin on the sides of the neck and capable of admitting a thin probe. Sometimes the place where they have been in childhood is marked throughout life by small round patches of white skin. These relics of the sea, these apparitions of the Fish, these sudden resurrections, are betrayals of man's pedigree. Men wonder at mummy-wheat germinating after a thousand years of dormancy. But here are ancient features bursting into life after unknown ages, and challenging modern science for a verdict on their affinities.

When the fish came ashore, its water-breathing apparatus was no longer of any use to it. At first it had to keep it on, for it took a long time to perfect the air-breathing apparatus which was to replace it. But when this was ready the problem was, what to do with the earlier organ? Nature is exceedingly economical, and could not throw all this mechanism away. In fact Nature almost never parts with any structure she has once made. What she does is to change it into something else. Conversely, Nature seldom makes anything new; her method of creation is to adapt something old. Now when Nature started out to manufacture ears, she made them out of the old breathing apparatus. She saw that if water could pass through a hole in the neck, sound could likewise, and she set to work upon the highest up of the five gill-slits and slowly elaborated it into a hearing organ.

There never had been an external ear in the world till this was done, or any good ear at all. Creatures which live in water do not seem to use hearing much, and the sound-waves in fishes are simply conveyed through the walls of the head to the internal ear without any definite mechanism. But as soon as land-life began, owing to the changed medium through which sound-waves must now be propagated, a more delicate instrument was required. And hence one of the first things attended to was the construction and improvement of the ear.

It has long been a growing certainty to Comparative Anatomy that the external and middle ear in Man are simply a development, an improved edition, of the first gill-cleft and its surrounding parts. The tympano-Eustachian

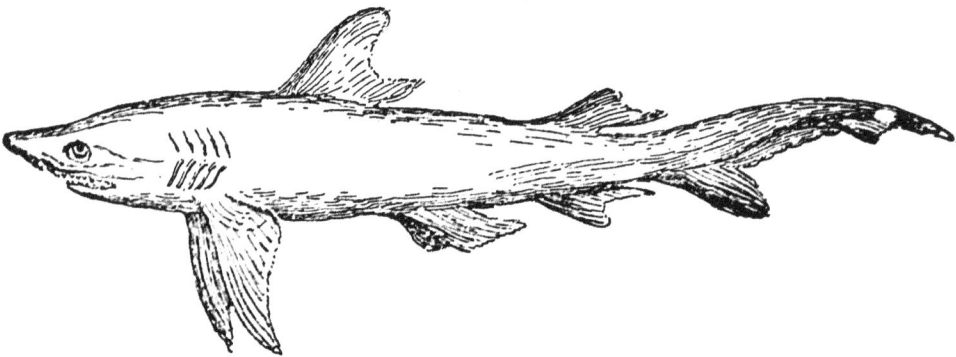

Adult shark (after Cuvier and Haeckel)

—from "Darwin and After Darwin."

passage is the homologue or counterpart of the spiracle, associated in the shark with the first gill-opening. Professor His of Leipsic has worked out the whole development in minute detail, and conclusively demonstrated the mode of origin of the external ear from the coalescence of six rounded tubercles surrounding the first branchial cleft at an early period of embryonic life. Haeckel's account of the process is as follows: "All the essential parts of the middle ear—the tympanic membrane, tympanic cavity, and Eustachian tube— develop from the first gill-opening with its surrounding parts, which in the Primitive Fishes (*Selachii*) remains throughout life as an open blow-hole, situated between the first and second gill-arches. In the embryos of higher Vertebrates it closes in the centre, the point of concrescence forming the tympanic membrane. The remaining outer part of the first gill-opening is the rudiment of the outer ear-canal. From the inner part originates the tympanic cavity, and further inward, the Eustachian tube. In connection with these, the three bonelets of the ear develop from the first two gill-arches; the hammer and anvil from the first, and the stirrup from the upper end of the second gill-arch. Finally as regards the external ear, the ear-shell (*concha auris*), and the outer ear-canal, leading from the shell to the tympanic membrane—these parts develop in the simplest way from the skin-covering which borders the outer orifice of the first gill-opening. At this point the ear-shell rises in the form of a circular fold of skin, in which cartilage and muscles afterwards form." [3]

Now bearing in mind this account of the origin of ears, an extraordinary circumstance confronts us. Ears are actually sometimes found bursting out *in human beings* half-way down the neck, in the exact position—namely along the line of the anterior border of the sterno-mastoid muscle—which the gill-slits would occupy if they still persisted. In some human families where the tendency to retain these special structures is strong, one member sometimes illustrates the abnormality by possessing the clefts alone, another has a cervical ear, while a third has both a cleft and an ear,—all these of course in addition to the ordinary ears. This cervical auricle has all the characters of the ordinary ear, "it contains yellow elastic cartilage, is skin-covered, and has muscle-fibre attached to it."[4]

Dr. Sutton further calls attention to the fact that on ancient statues of fauns and satyrs cervical auricles are sometimes found, and he figures the head of a satyr from the British Museum, carved long before the days of anatomy, where a sessile ear on the neck is most distinct. A still better illustration may be seen in the Art Museum at Boston on a full-sized cast of a faun belonging to the later Greek period; and there are other examples in the same

Marble head of satyr, in Munich, showing cervical auricles.

building. One interest of these neck-ears in statues is that they are not as a rule modelled after the human ear but taken from the cervical ear of the goat, from which the general idea of the faun was derived. This shows that neck-ears were common on the goats of that period—as they are on goats to this day—but the sculptor would hardly have had the daring to introduce this feature in the human subject unless he had been aware that pathological facts encouraged him. The occurrence of these ears in goats is no more than one would ex-pect. Indeed one would look for them not only in Man, but in all the Mammalia, for so far as their bodies are concerned all the higher animals are near relations. Observations on vestigial structures in animals are sadly wanting; but they are certainly found in the horse, pig, sheep, and others.

That the human ear was not always the squat and degenerate instrument it is at present may be seen by a critical glance at its structure. Mr. Darwin records how a celebrated sculptor called his attention to a little peculiarity in the external ear, which he had often noticed both in men and women. "The peculiarity consists in a little blunt point, projecting from the inwardly folded margin or helix. When present, it is developed at birth, and according to Professor Ludwig Meyer, more frequently in man than in woman. The helix obviously consists of the extreme margin of the ear folded inwards; and the folding appears to be in some manner connected with the whole external ear being permanently pressed backwards. In many monkeys who do not stand high in the order, as baboons and some species of macacus, the upper portion of the ear is slightly pointed, and the margin is not at all folded inwards; but if the margin were to be thus folded, a slight point would necessarily project towards the centre."[5]

Here then, in this discovery of the lost tip of the ancestral ear, is further and visible advertisement of man's Descent, a surviving symbol of the stirring times and dangerous days of his animal youth. It is difficult to imagine any other theory than that of Descent which could account for all these facts. That

evolution should leave such clues lying about is at least an instance of its can-
dor.

But this does not exhaust the betrayals of this most confiding organ. If we turn from the outward ear to the muscular apparatus for working it, fresh traces of its animal career are brought to light. The erection of the ear, in or-der to catch sound better, is a power possessed by almost all mammals, and the attached muscles are large and greatly developed in all but domesticated forms. This same apparatus, though he makes no use of it whatever, is still attached to the ears of Man. It is so long since he relied on the warnings of hearing, that by a well-known law the muscles have fallen into disuse and at-rophied. In many cases, however, the power of twitching the ear is not wholly lost, and every school-boy can point to some one in his class who retains the capacity and is apt to revive it in irrelevant circumstances.

One might run over all the other organs of the human body and show their affinities with animal structures and an animal past. The twitching of the ear, for instance, suggests another obsolete or obsolescent power—the power, or rather the set of powers, for twitching the skin, especially the skin of the scalp and forehead by which we raise the eyebrows. Sub-cutaneous muscles for shaking off flies from the skin, or for erecting the hair of the scalp, are common among quadrupeds, and these are represented in the human sub-ject by the still functioning muscles of the forehead, and occasionally of the

Horned sheep and goat, showing cervical auricles .

—from "Evolution and Disease," J. Bland-Sutton

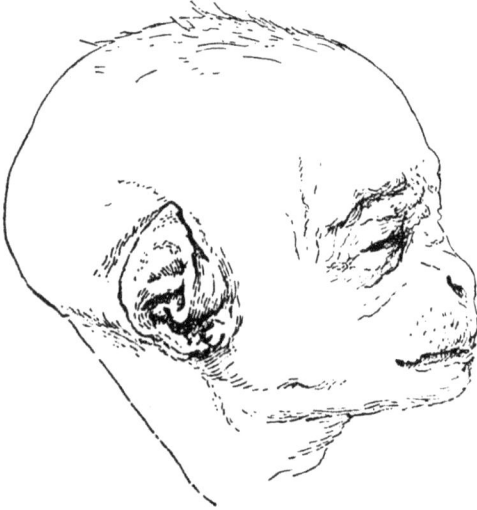

Form of the ear in baby outang.

—from "Darwin and After Darwin"

head itself. Everyone has met persons who possess the power of moving the whole scalp to and fro, and the muscular apparatus for effecting it is identical with what is normally found in some of the Quadrumana.

Another typical vestigial structure is the *plica semi-lunaris*, the remnant of the nictitating membrane characteristic of nearly the whole vertebrate sub-kingdom. This membrane is a semi-transparent curtain which can be drawn rapidly across the external surface of the eye for the purpose of sweeping it clean. In birds it is extremely common, but it also exists in fish, mammals, and all the other vertebrates. Where it is not found of any functional value it is almost always represented by vestiges of some kind. In Man all that is left of it is a little piece of the curtain draped at the side of the eye.

When one passes from the head to the other extremity of the human body one comes upon a somewhat unexpected but very pronounced characteristic— the relic of the tail, and not only of the tail, but of muscles for wagging it. Everyone who first sees a human skeleton is amazed at this discovery. At the end of the vertebral column, curling faintly outward in suggestive fashion, are three, four, and occasionally five vertebrae forming the coccyx, a true rudimentary tail. In the adult this is always concealed beneath the skin, but in the embryo, both in man and ape, at an early stage it is much longer than the limbs. What is decisive as to its true nature, however, is that even in the embryo of man the muscles for wagging it are still found. In the grown-up human being these muscles are represented by bands of fibrous tissue, but cases are known where the actual muscles persist through life. That a distinct external tail should not be still found in Man may seem disappointing to the evolutionist. But the want of a tail argues more for the theory of Evolution than its presence would have done. It would have been contrary to the Theory of Descent had he possessed a longer tail. For all the anthropoids most allied to Man have long since also parted with theirs.

It was formerly held that the entire animal creation had contributed something to the anatomy of Man, that as Serres expressed it "Human Organo-genesis is a condensed Comparative Anatomy." But though Man has not such a monopoly of the past as is here inferred—other types having here and there emerged and developed along lines of their own—it is certain that the materials for his body have been brought together from an unknown multitude of lowlier forms of life.

Those who know the Cathedral of St. Mark's will remember how this noblest of the Stones of Venice owes its greatness to the patient hands of centuries and centuries of workers, how every quarter of the globe has been spoiled of its treasures to dignify this single shrine. But he who ponders over the more ancient temple of the human body will find imagination fail him as he tries to think from what remote and mingled sources, from what lands, seas, climates, atmospheres, its various parts have been called together, and by what innumerable contributory creatures, swimming, creeping, flying, climbing, each of its several members was wrought and perfected. What ancient chisel first sculptured the rounded columns of the limbs? What dead hands built the cupola of the brain, and from what older ruins were the scattered pieces of its mosaic-work brought? Who fixed the windows in its upper walls? What forgotten looms wove its tapestries and draperies? What winds and weathers wrought the strength into its buttresses? What ocean-beds and forest glades worked up the colors? What Love and Terror and Night called forth the Music? And what Life and Death and Pain and Struggle put all together in the noiseless workshop of the past and removed each worker silently when its task was done? How these things came to be Biology is one long record.

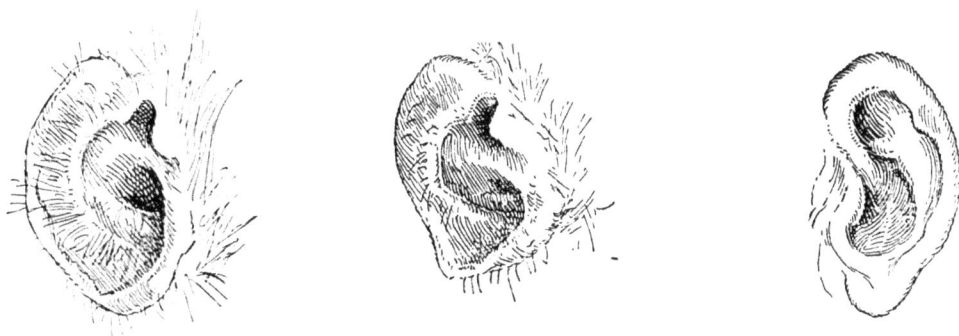

Ear of Barbary ape, chimpanzee, and man, showing vestigial characters of the human ear.

—from "Darwin and After Darwin"

The architects and builders of this mighty temple are not anonymous. Their names, and the work they did, are graven forever on the walls and arches of the Human Embryo. For this is a volume of that Book in which Man's members were written, which in continuance were fashioned, when as yet there was none of them.

[1] N.B.—They appear as "clefts," marking not the adult fish, but the embryo at the corresponding stage.
[2] "Evolution and Disease," p. 81.
[3] Haeckel: "Evolution of Man," vol. ii, p. 269.
[4] Sutton: "Evolution and Disease."
[5] "Descent of Man," p. 15.

Foods in the Year 2000

Professor Berthelot's Theory That Chemistry Will Displace Agriculture.

By Henry J. W. Dam

September 1894

"**W**HAT will the man of the future eat?" The answer to this question has been undertaken, not by an imaginative writer, but by one of the greatest of living men of science, Professor Berthelot of Paris; and it may be said at once that, but for his scientific eminence and the undeniable facts upon which he bases his forecast, it would pass the limits of human belief. The epicure of the future is to dine upon artificial meat, artificial flour, and artificial vegetables; drink artificial wines and liquors, and round off his repast with an artificial tobacco beside which the natural tobacco of the present will seem poor indeed.

WHEAT FIELDS AND CORN FIELDS TO DISAPPEAR.

Wheat fields and corn fields are to disappear from the face of the earth, because flour and meal will no longer be grown, but made. Herds of cattle, flocks of sheep, and droves of swine will cease to be bred, because beef and mutton and pork will be manufactured direct from their elements. Fruit and flowers will doubtless continue to be grown as cheap decorative luxuries, but no longer as necessities of food or ornament. There will be in the great air trains of the future no grain or cattle or coal cars, because the fundamental food elements will exist everywhere and require no transportation. Coal will no longer be dug, except perhaps with the object of transforming it into bread or meat. The engines of the great food factories will be driven, not by artificial combustion, but by the underlying heat of the globe.

In order to clearly conceive these impending changes, it must be remem-

Professor Berthelot

bered that milk, eggs, flour, meat, and, indeed, all edibles, consist almost entirely (the percentage of other elements is very small) of carbon, hydrogen, oxygen, and nitrogen. Oxygen and hydrogen are the two gases which, when combined, form water. Oxygen and nitrogen mixed are the air we breathe. Carbon forms the charcoal of wood, is the main constituent of coal, and as carbonic acid gas in the air is the chief food of the vegetable world. These four elements, universally existing, are destined to furnish all the food now grown by nature, through the rapid and steady advance of synthetic chemistry.

Synthetic chemistry is the special science which takes the elements of a given compound, and induces them to combine and form that compound. It is the reverse of analytic chemistry, which takes a given compound, and dissociates and isolates its elements. Analytic chemistry would separate water into oxygen and hydrogen, and synthetic chemistry would take oxygen and hydrogen, mix them, put a match to the mixture, and thus form water. For many years past synthetic chemistry has had an eager eye upon foodmaking. It has already progressed so far that several great agricultural industries have been destroyed by its advancement, compounds which were once obtained by plant growth in the fields being now entirely furnished by chemical laboratories and direct manufacture. In fact, the clear evidence of the present leads quite logically to the conclusion that at some more or less distant period in the future, synthetic chemistry will destroy all the great agricultural industries, and put to new uses the grain fields and cattle ranges of to-day.

PROFESSOR BERTHELOT.

No man is more entitled to act as a prophet in this field than Professor Berthelot. If not the father, he is certainly the foster-father, of synthetic chemistry as a special science, and for nearly fifty years he has been one of the leaders of the scientific army in the invasion of strange regions. In every way open to a grateful nation, France has loaded him with honors. He is a member and perpetual secretary of the Academy of Sciences, a member of the Institute, and a grand officer of the Legion of Honor. He is president of the Superior Council of Public Instruction, president of the Committee on Explosives, and in 1870 was president of the Committee on the Defence of Paris. As a cabinet minister he has had occasion and excellent opportunity to study the people, and as a life-long chemist he has enjoyed the best opportunities for considering the industrial changes which affect their condition. Many of the manufacturing improvements which have enriched France have been due, directly or indirectly, to his own chemical researches. Consequently, any predictions he ventures in his chosen field have the highest value, and I was particularly glad of an interview which he was good enough to accord me lately.

Professor Berthelot occupies in Paris a residential suite of apartments in the Institute of France. This is a great stretch of old brown stone buildings on the Quai Conti, with bare and barren courtyards paved with many square feet of gray stone blocks. The sense of coldness in the environment that you have after you have traversed two of these courtyards to the last doorway on the right, is dissipated by the cheery smile of a stout Breton sewing woman, who ushers you without delay through a long, dark corridor to a small, dark study at the end. Here, surrounded by books, which cover the walls on all four sides, sits at his desk the Professor. His slender figure, clad in professional black, is somewhat bent by the deep study which has made his fame, but otherwise his sixty-seven years sit lightly upon him. His greeting is grave, but entirely courteous and sympathetic, an intelligent curiosity concerning the field of research to which he has devoted his life being all that is required to arouse his interest and unlock his store of strange and interesting facts. The interview is had pursuant to an appointment, and he plunges at once into the subject, referring to his address of April 5th before the Society of Chemical and Mechanical Industries.

"That address," he says, "was in the nature of an after-dinner speech rather than a scientific pronouncement. We do not use the dryer language of science upon festive occasions. I was speaking, however, to an association of chemists, and I believe that all I predicted upon that occasion will, in the process of time, say the year 2000, be actually or approximately the existing state of affairs. I said that new sources of mechanical energy would largely replace the present use of coal, and that a great proportion of our staple foods, which we now obtain by natural growth, would be manufactured direct, through the advance of synthetic chemistry, from their constituent elements, carbon, hydrogen, oxygen, and nitrogen. I not only believe this, but I am unable to doubt it. The tendency of our present progress is along an easily discerned line, and can lead to only one end."

"Do you mean to predict that all our milk, eggs, meat, and flour will in the future be made in factories?"

A TABLET OF FACTORY-MADE BEEF-STEAK.

"Why not, if it proves cheaper and better to make the same materials than to grow them? The first steps, and you know that it is always the first step that costs, have already been taken. It is many years, you must remember, since I first succeeded in making fat direct from its elements. I do not say that we shall give you artificial beefsteaks at once, nor do I say that we shall ever give you the beefsteak as we now obtain and cook it. We shall give you the same identical food, however, chemically, digestively, and nutritively speaking. Its form will differ, because it will probably be a tablet. But it will be a tablet of any color and shape that is desired, and will, I think, entirely satisfy the epicurean senses of the future; for you must remember that the beefsteak of to-day is not the most perfect of pictures either in color or composition."

This declaration from so high an authority was somewhat staggering. It was an unexpected blow at a tender (usually tender) and long-loved household idol. The common or garden beefsteak suddenly took upon itself a poetry and a pathos in the mind which could only have been born of its prospective superannuation. The idea of glass cows and brass beefsteak-machines, however scientific, carried a certain shock which was scarcely modified by the hope that the beefsteak of the future might, could, and would escape ever being tough.

"To comprehend what I mean by the tendency of the time," continued Professor Berthelot, "you must consider the long evolution which has characterized the development of foods and the major part which chemistry has played therein. The point is, that from the earliest time we have steadily increased our reliance upon chemistry in food production, and just as steadily diminished our reliance upon nature. Primitive man ate his food and vegetables raw. When he began to cook, when he first used fire, chemistry made its first intrusion upon the sphere of nature. To-day the fire in the open air has been replaced by the kitchen. Every cooking utensil now used represents some one of the chemical arts. Stoves, saucepans, and pottery are the results of chemical industries. So also modern cookery uses an indefinite number of compounds—food compounds—which, like sugar for instance, have been subjected to a more or less complex chemical treatment in their journey from the field in which they grew to the kitchen in which they are used. The ultimate result is clear. Chemistry has furnished the utensils, it has prepared the foods, and now it only remains for chemistry to make the foods themselves, which, indeed, it has already begun to do."

Before proceeding to describe what synthetic chemistry has already done in this direction, the Professor said, by way of preface:

"There is a distinction which I would like to make at this point between the laboratory stage and the commercial stage of any given discovery in food-making. From the scientific standpoint the laboratory result is the important one. As you and all the world know, the commercial result follows inevitably in time. Once science has declared that a desired end is attainable, the genius of invention fastens upon the problem, and the commercial production of the result slowly attains perfection by gradually improved processes at less and less cost. Take aluminium, for instance. Once a very expensive metal, its steadily decreased cost in production is bringing it within the reach of all. The use of sugar is universal. Sugars have recently been made in the laboratory. Commerce has now taken up the question, and I see that an invention has recently been patented by which sugar is to be made upon a commercial scale from two gases, at something like one cent per pound. As to whether or not the gentlemen who own the process can do what the inventor claims, it is neither my province nor my desire to express an opinion. It may be that the commercial synthetic manufacture of sugar is a more difficult task than they imagine. I have not the slightest doubt, however, that sugar will eventually be manufactured on the largest scale synthetically, and that the culture of the sugar-cane and the beet-root will be abandoned because they have ceased to pay. Look at alizarin. There is one result of the same kind that synthetic chemistry has already brought about."

"What is alizarin?"

A GREAT AGRICULTURAL INDUSTRY ALREADY DESTROYED BY THE CHEMIST.

"Alizarin is a compound whose synthetic manufacture by chemists has destroyed a great agricultural industry. It is the essential commercial principle of the madder root, which was once used in dyeing wherever dyeing was carried on. The madder root was grown to an enormous extent in Persia, India, and the Levant, and spread from there to Spain, Holland, and the Rhine provinces. Continental Europe used it in enormous quantities, and twenty years ago its annual import into England was valued at six million two hundred and fifty thousand dollars. The discovery was made, however, that alizarin could be manufactured synthetically, and the artificial production of it has so far supplanted the natural that the madder fields, so far as Europe is concerned, have practically ceased to exist. So with indigo. The chemists have now succeeded in making pure indigo direct from its elements, and it will soon be a commercial product. Then the indigo fields, like the madder fields, will be abandoned, industrial laboratories having usurped their place."

So far as dye stuffs were concerned, the intervention of chemistry seemed not so unnatural. When it came to tobacco and tea and coffee, however, synthetic chemistry appeared to be getting nearer home, invading the family circle, so to say.

"Tea and coffee could now be made artificially," continued the Professor, "if the necessity should arise, or if the commercial opportunity, through the necessary supplementary mechanical inventions, had been reached. The essential principle of both tea and coffee is the same compound. The difference of name between theine and caffeine has arisen from the sources from which they were obtained. They are chemically identical in constitution, and their essence has often been made synthetically. The scale of manufacture, or synthetic ladder, is as follows:

> Carbon and oxygen make carbonic oxide.
> Carbonic oxide and chlorine make carbonyl chloride.
> Carbonyl chloride and ammonia make urea, whence uric acid.
> Uric acid transforms into xanthine.
> Xanthine yields theobromine.
> Theobromine yields theine or caffeine.

Theobromine, you remember, is the essential principle of cocoa. Thus, you see, synthetic chemistry is getting ready to furnish, from its laboratories, the three great non-alcoholic beverages in general use. The tea plants, coffee shrubs, and cocoa trees must some day follow the lead of madder and indigo."

"And what about tobacco?"

EXCELLENT TOBACCO OUT OF COAL TAR.

"The essential principle of tobacco, as you know, is nicotine. We have obtained pure nicotine, whose chemical constitution is perfectly understood, by treating salomine, a natural glucoside, with hydrogen. Synthetic chemistry has not made nicotine directly as yet, but it has very nearly reached it, and the laboratory manufacture of nicotine may fairly be expected at any time. Conine, the poisonous principle of hemlock, has been made synthetically, and it is so close in its constitution to nicotine, and so clearly of the same class, that only its transformation into nicotine remains to be mastered, a problem which is not very difficult when compared with others which have been solved. The parent compound from which the nicotine of commerce will be made, exists largely in coal tar."

"You believe, then, that all our tobaccos will some day be made artificially?"

"To as great an extent as appears desirable. The choicer growths, with their individual characteristics from individual circumstances of growth, will be longest cultivated. The tobacco leaf is simply so much dried vegetable matter, in which nicotine is naturally stored. Chemistry will first make the nicotine, and impregnate any desirable leaf with it to any degree of strength. Later on, if necessary, it will also make the leaf. In some directions it is not difficult to improve upon nature, and the best chemical medium for carrying nicotine might easily prove superior to the natural."

Having weakly permitted his beefsteak to be carried by storm, the writer was all the more inclined to defend his tobacco. "But, surely," said he, "there is something more in fine tobacco than merely nicotine and vegetable fibre."

"Precisely. Leaving aside what the manufacturers may add, there are delicate flavoring oils, which chemistry will also create. Vanilla, a flavoring compound of very general use, has always been obtained, until recently, from the tonka bean. Now artificial vanillin, in the same compound made chemically, threatens to drive the natural vanilla out of the European market, and will doubtless succeed in doing so as its manufacture is perfected. In fact, some of the chocolate and confectionery manufacturers are already taking it up. All the essential oils will eventually be made direct. Vanillin is very near in its chemical constitution to the aromatic, the distinctive, principle of cloves and allspice. Artificial cloves and allspice will therefore probably come next. Flower perfumes, too, have been fully analyzed, and in time will be largely synthesized. One of them, meadow-sweet, is being largely compounded and sold. There are consequently no virtues in the natural tobacco which are likely to be missed in the artificial. In fact, the contrary state of affairs is more probable."

With our tobacco prospectively obtained from coal tar, and our flower perfumes made without flowers, the sphere of synthesis was decidedly broadening. Professor Berthelot, however, made it broader, touching upon an important law of which he himself was the discoverer.

USEFUL COMPOUNDS UNKNOWN TO NATURE IN PROSPECT.

"Perhaps the greatest importance, and certainly the profoundest charm, in the study of synthetic chemistry," said he, " is the certain evidence which it offers of the discovery and manufacture of many compounds now entirely unknown, whose effect upon human health, human life, and human happiness no one can possibly conjecture. A hundred years ago, for instance, who could have foretold photography or the telephone? To understand the prospect before us, you must recall the familiar laws of atomic proportion which characterize the combination of different elements in mineral chemistry. Among the metals of a given class, the combinations with oxygen, for example, follow the same atomic law, and, by means of this law, compounds which had not been met with by chemists were from time to time looked for and created. In other words, the acting law of atomic proportions having been discovered, the chemist knew what to expect in a given class of metals, and wrote the formula, in many cases, before he obtained the compound.

"Now, in mineral chemistry the operations of this law are limited. In organic chemistry they are limitless. When I found that the same general laws acted uniformly throughout the measureless field of organic chemistry, I saw that the number of compounds to be dealt with in this way was numberless, infinite in extent. How many compounds each having, like nicotine or theine, its own distinctive peculiarities, remain to be discovered, it is simply beyond our power to imagine. Analytical chemistry has dealt and could deal only with what it found in nature. Synthetic chemistry, armed with the elements and with the knowledge of the laws by which classes of complex compounds combine with other classes of complex compounds, will go on and on, developing new fixed compounds not yet met with in nature, whose influence upon human life, as I have said, no mind can possibly foretell. The science of chemistry grew out of the search for two things, the elixir of life and the philosopher's stone. These two will never be discovered; but, that other compounds almost as wonderful in their way may be, it would be dogmatic and rash to deny. It is a great, untraversed country, a field of exploration for chemical students for centuries to come."

In illustration of this point, as well of his own methods of study, he continued:

"I passed hydrogen over carbon at a white heat, and by aid of the electric spark obtained a combination, the result being a familiar gas, acetylene.

"I found that acetylene would take up another atom of hydrogen in the nascent state, and this gave me marsh gas and ethylene.

"I found that ethylene in the presence of water could be made to combine with the oxygen and hydrogen of the latter, thereby forming ordinary alcohol, and that marsh gas in the same way formed methylic alcohol.

"Combining the acetylene with free oxygen in the nascent state, I obtained oxalic acid.

"Combining the acetylene with free nitrogen by aid of the electric spark, I obtained cyanhydric acid.

"Combining the acetylene with oxygen in the presence of water and an alkali, I obtained acetic acid.

"I also found that under certain conditions the acetylene could be transformed directly into benzine. Here, then, we have seven familiar compounds of wide utility, acetylene, marsh gas, alcohol, oxalic acid, acetic acid, cyanhydric acid, and benzine, to say nothing of many others which I might mention, obtained from these elements direct. Now, imagine for a moment the enormous number of organic compounds into the constitution of which, according to regularly acting laws, these seven compounds enter. There are six different families of alcohols alone, and each one of these families embraces a greater or less number of special alcohols. Here is a great list of alcohols, each one representing in combination an indefinite class of alkaloids. Some of these alkaloids are known, many more can be conjectured, and what may not develop from them when they are studied and tested in their relation to life and the arts? Over the whole field of organic chemistry the mystery of possibilities extends. Its combinations and intercombinations are so limitless that we can only work on regularly to ends that it is impossible to foresee. A discovery may arise at any time which will have an incalculable effect upon the happiness and welfare of man."

"What do you think of Tyndall's dogma, that from the non-living the living can never be obtained?"

"We do not know enough about the question to dogmatize," said Professor Berthelot, simply. "We shall see what we shall see."

Outside of the products named, the proved results of synthetic chemistry appeared to be already numerous. The oil of bitter almonds is now being made direct commercially, and so also is the oil of mustard. Mustard made from the oil of mustard is preferred for use as an irritant by many physicians, in consequence of its purity, which is perfect; whereas the natural mustard contains other compounds not entirely desirable in this connection. Salicylic acid, tar-

taric acid, the acid of unripe grapes, and citric acid, the sour principle of lemons and other fruits, are made direct. Artificial turpentine is being actively sought after, and from it chemists expect to obtain artificial caoutchouc. Long before the promised failure of the rubber trees to supply the demands of commerce, synthetic rubber will, in all probability, have filled the void. From a review of what synthetic chemistry has already done, the Professor passed to the subject of what it may reasonably be expected to do. He said:

ASS'S MILK, GOAT'S MILK, AND COW'S MILK ALL FROM ONE LABORATORY.

"The production of food stuffs upon a commercial scale by synthetic chemistry will naturally depend upon two things: the cheapness of production, and the quality of the result. Take artificial butter as an instance. Twenty years ago in this country the idea was conceived of making butter from beef fat. This was an intermediate synthesis, which consisted in extracting the oleopalmitine from the fat by melting, cooling, and pressure. The extract was then treated with milk, churned, and colored as the dairymen colored natural butter. The growth of the oleomargarine industry has been extensive, and its manufacture now takes place on a large scale. The best artificial butter approaches so closely to real butter that the difference is not very great. In the same way, whenever synthetic chemistry is ready and the commercial conditions have been met, artificial chemical food will infringe upon the sphere of the natural in other directions. Nature, however, produces very cheaply; and no man will desire to lose money for the pleasure of making chemical food. There is no reason, nevertheless, since we are making artificial butter, why we should not before long make artificial milk. Milk consists of say 3.50 per cent of milk fats—olein, stearine, butyrine, palmitine, and others—3.98 per cent, of caseine, four per cent of milk-sugar, and 86.87 per cent, of water, with traces of other substances which have been determined. It will not be a very difficult problem for synthetic chemistry to mingle these constituents in these proportions, and make a milk that will as nearly approach natural milk, in meeting the demands and desires of the public, as artificial butter approaches natural butter. So, too, the variation of proportions would be easy; and ass's milk, goat's milk, or any other milk desired, could be furnished from the same laboratory as easily as cow's milk. The only necessity is, that we shall be able to make all the solid constituents mentioned, and this is simply a matter of time. In short, milk factories may be looked for just as soon as the constituents can be directly and cheaply obtained."

"But will such milk be as healthful as that produced naturally?"

"There is no reason why it should not be. With the vital action and vital machinery by which the cow produces the milk, chemistry has nothing to do. It is a question of physiology. When the milk has left the cow, however, it is merely a chemical compound, and with it physiology has nothing to do. As I said, the fats I have already made direct. The milk-sugar, too, has been made. When we come to the caseine, however, and, with it, to starch, meat, and albumen, we come into a set of very complex chemical problems. Still, they are merely chemical problems, and as such are subject to study and solution in the future, just as we have seen equally difficult problems met and solved in the past. The mass of animal tissues are constituted of certain nitrogenous compounds which play an equally important rôle in the development of vegetable tissues. These compounds are very complex, nearly always fixed and uncrystallizable, and easily affected by re-agents. Some are soluble and some insoluble, but most of the former become insoluble by coagulation in water, through heat or through the action of acids. Such are albumen, fibrine, caseine, syntonine, osseine, chondrine, glutine, chitine, etc. To make meat we must make these compounds, or so many of them as are necessary. That chemistry will some day be able to make them I cannot doubt. That at some time in the future artificial meat will infringe upon the domain of natural meat, as artificial butter has upon that of natural butter, is only to be reasonably expected. So with the vegetables. A potato consists of, say, 81.844 Per cent, of starch, 13.030 per cent, of water, 2.313 per cent, of nitrogenous matter, 1.13 per cent, of woody fibre, and minute proportions of fat and mineral constituents. When we are able to make starch direct, what will hinder us from making a potato? And what is to prevent us, once we have gained the mastery, from making better milk, better meat, and better potatoes, at any season of the year, than those which nature gives us, more or less afflicted, as these are, with impurities and additions, and produced only at the periods in which her laboratories are kept open for the public good?

STEAM TO BE PIPED FROM THE CENTRE OF THE EARTH.

"Time is not an element in these speculations, because all the future is before us, and the line of march is marked out. Great changes, however, which will cheapen the cost of producing these results, will come from cheaper and simpler sources of mechanical energy than those now used. Herein lies the fundamental problem of all the industries, to discover sources of energy which are inexhaustible and which will renew themselves without effort on our part.

Nature has given us these, ready for our use, but as yet we have accepted only a very small portion of her gift. Evolution has long acted in this direction also, and must continue to act. We have seen the force of human hands largely replaced by that of steam, that is to say, by chemical energy borrowed from the combustion of coal. Coal, however, is laboriously extracted from the bowels of the earth. The time is coming when, by methods already foreseen and unnecessary to describe, we shall store and make use of the heat of the sun. But greater, far greater in importance than this, will be the ultimate and widespread use of the central heat of our globe. The incessant advances of science give us a sure basis upon which to expect a limitless amount of energy drawn from this source. It will suffice, to utilize the central heat, to sink pits from three thousand to four thousand metres (three thousand three hundred to four thousand four hundred yards) in depth, and this is a problem in engineering quite within the powers of the engineers of the present day, to say nothing of the engineers of the future. At this depth we should find great heat, constant and unvarying, the heat which is the source of all energy and all life."

The writer could personally vouch for this. He recalled having his back scalded by dripping water in one of the three thousand two hundred foot levels of the Comstock Lode.

"At these depths," said Professor Berthelot, "we may easily tap superheated steam under pressure which can be used to drive machinery direct from the top of the shaft. That, however, is merely a detail. We shall have in these pits the cheapest of furnaces, because we can have them at any degree of heat, never failing and never needing fuel or renewal. They will be at some distance from our engines, to be sure, but that will be no difficulty. Into them we can introduce water, if necessary, convert it into superheated steam at the bottom, and use it on the surface. More than this, the advance of thermoelectric science is certain, once the inventions are needed, to supply us with another and perhaps more convenient means of turning this heat into force and using it for mechanical purposes at a cost, after the plant is constructed, which will be no more than the wear and tear. We shall thus have a source of energy which costs nothing, whose extent is indefinite, which is incessantly renewed, and whose diminution through centuries will be quite imperceptible. And this will be force which will be available everywhere, all over the globe, and equally the blessing, with the property which results from it, of all nations and all peoples which seek its use. Given such sources of energy, the artificial production of food will be a much simpler problem, and will more rapidly fall into the hands of chemistry. The hard preliminary work is done. The synthesis of the fats and oils I myself accomplished years ago. That of the sugars and carbo-hydrates is the study of the present time, and that of the nitrogenous compounds is not far off. Carbon from carbonic acid, oxygen and hydrogen from water, and nitrogen from the air will be a source of food for all the world. What the animals and vegetables have produced through the energy of nature, we shall produce as well, if not better, by our study of nature's laws. Strange though it may seem, the day will come when man will sit down to dine from his toothsome tablet of nitrogenous matter, his portions of savory fat, his balls of starchy compounds, his casterful of aromatic spices, and his bottles of wine or spirits which have all been economically manufactured in his own factories, independent of irregular seasons, unaffected by frost, and free from the microbes with which over-generous nature sometimes modifies the value of her gifts."

"And all this will be due to chemistry?"

A SYNTHETIC ARCADIA.

"To chemistry and her sister science, physics. If one chooses to base dreams, prophetic fancies, upon the facts of the present, one may dream of alterations in the present conditions of human life so great as to be beyond our contemporary conception. One can foresee the disappearance of the beasts from our fields, because horses will no longer be used for traction or cattle for food. The countless acres now given over to growing grain and producing vines will be agricultural antiquities, which will have passed out of the memory of men. The equal distribution of natural food materials will have done away with protectionism, with custom-houses, with national frontiers kept wet with human blood. Men will have grown too wise for war, and war's necessity will have ceased to be. The air will be filled with aerial motors flying by forces borrowed from chemistry. Distances will diminish, and the distinction between fertile and non-fertile regions, from the causes named, will largely have passed away. It may even transpire that deserts now uninhabited may be made to blossom, and be sought after as great seats of population in preference to the alluvial plains and rich valleys, soils fat with putrefaction, which constitute the great agricultural and popular centres of to-day."

"And man?"

"Man should grow in sweetness and nobility, because he will have done with war, with existence based upon the slaughter of beasts. Perhaps—this is only a dream, remember—synthetic chemistry, or something that we might call spiritual chemistry, will develop means to as profoundly alter man's moral nature as material chemistry will change the conditions of his environment. There is no fear that art, beauty, and the charm of human existence are destined to disappear. If the surface of the earth ceases to be divided, and I may say disfigured, by the geometrical devices of agriculture, it will regain its natural verdure of woods and flowers. Man, becoming familiar with the principles and responsibilities of self-government, will be more easily governed. The favored portions of the earth will become vast gardens, in which the human race will dwell amid a peace, a luxury, and an abundance recalling the Golden Age of legendary lore.

"These are dreams, of course," added the Professor in conclusion, "but science may surely be permitted to dream sometimes. If it were not for our dreams, where would be our impulse to progress?"

The Search for the Absolute Zero

By Henry J. W. Dam

November 1894

THE search for the zero of absolute temperature is being rapidly pushed forward. It is one of the strangest and most important quests ever undertaken by science, and its attainment will have an effect upon our general knowledge of the universe and of matter, for instance, compared to which the results of the discovery of the North Pole will be trifling. The temperature of the absolute zero is supposed to be the lowest point of cold existing or possible in the universe. It is the supposed starting point of that molecular motion which we call heat. To attain it would give us the basis for a new and absolute thermometer, which would in itself be an enormous advantage in many branches of natural science, among which physics, chemistry, and astronomy would reap the greatest benefit. All our present measurements of heat are relative. A Fahrenheit thermometer merely marks the height in a tube attained by a column of mercury at the temperature of melting ice, and the height it attains at the temperature of boiling water. Between these two points the tube has been arbitrarily divided, the expansion of the mercury by heat being uniform, into one hundred and eighty equal parts, called degrees. The division is also continued below the freezing point of water, and thirty-two more degrees are marked off, creating an arbitrary zero. In the centigrade thermometer the zero is assumed at the freezing point of water, and 100° are marked off between that and the boiling point. In the Réaumur thermometer the zero is at the freezing point of water, and 80° are marked off on the tube between that and the boiling point. Consequently, under a given atmospheric pressure the boiling point of water is 212° Fahrenheit, 100° centigrade, and 80° Réaumur. In other words, our measure of heat in general use in all laboratories and observatories has nothing absolute about it, but is merely a convenient means of comparing the heat of any object or place with the effects of heat upon water. To discover the absolute

zero and to make an absolute thermometer would change all this. It
would set the science of heat upon an
exact basis, and enable us to state
how hot, absolutely and not comparatively, any object or place in nature
might actually be.

The zero of absolute temperature
has long been indicated as a mysterious and important point in two ways.
The first is the contraction of gases,
which in all known gases operates
uniformly as the temperature is lowered. As long as they retain the gaseous state, gases shrink in volume so

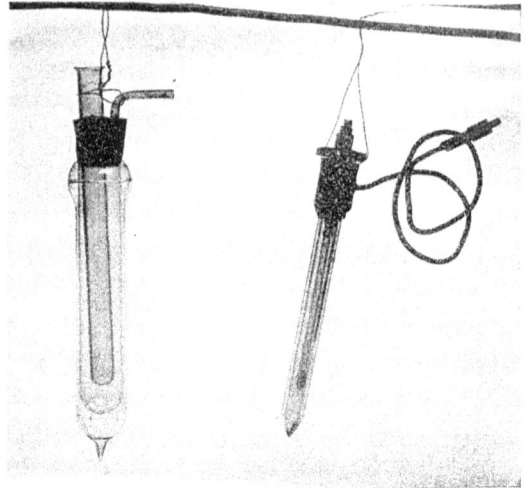

No. 1.   No. 2.

No. 1 new and No. 2 old type of tube used by
Professor Dewar in liquefying oxygen.

uniformly with each added degree of cold that an exact, unvarying line of diminishing volume is established. This line is as unvarying as the pointing of
the needle to the North Pole. It cannot be explained any more than the action
of the needle can be explained. As every gas is cooled, however, degree by degree, it points unerringly, by the law of diminishing proportions, to a point at
which its volume would be nothing. If the shrinkage continued, since the proportion of loss of volume never varies, the gas would shrink to nothingness. It
could not do so, of course; and all gases, sooner or later, fall out of the line by
becoming liquid, when the law ceases to operate, and the proportion of contraction in volume ceases to be the same. As long as they remain gases, however—and the law is precisely the same in all gases—they mechanically point
their figurative fingers in one direction, and all these figurative fingers indicate a point which is 461° below the zero of the Fahrenheit thermometer.

In a similar way this point is also indicated by all the pure metals. At ordinary temperatures the power of the pure metals to conduct electricity varies
exceedingly. Copper, iron, platinum, and lead have very different capacities in
this regard. As they are cooled, however, a change takes place in all. The resistance to the passage of electricity decreases. The poorest conductors at ordinary temperatures are those which offer most resistance to the electrical current. Under increasing cold these become better conductors rapidly. The line
of alteration in electrical resistance, as the temperature goes downward, is not
alike in any two. But the lines of several metals converge; they come closer
and closer together as the temperature approaches, say, 328° below zero Fahr-

enheit. And these lines of convergence point, in the same strange way as the gases, to the same point, 461° below zero Fahrenheit, as the point at which they would all meet. In other words, there is a point at which the electrical conductivity of all pure metals would be the same.

These are the two processes which have pointed out the zero of absolute temperature as a point, of great importance. Moreover, as temperatures go downward, matter seems to die, to lose all its characteristics as we know them at ordinary temperatures. No two elements combine at ordinary temperatures with more powerful and flaming chemical energy than phosphorus and oxygen. Their chemical affinity is very great, and phosphorus burns in oxygen gas with great and powerful rapidity. But when a bit of phosphorus is put in a vessel of liquid oxygen, nothing can induce the two to combine. The phosphorus floats about in the liquid oxygen like a chip of wood in water. Chemical action, the force of chemical affinity, has ceased. This is true of all chemical action at this temperature; and so it is known that at 461° below zero Fahrenheit all matter, chemically speaking, must be dead. In fact, no greater cold than the temperature of the absolute zero can be conceived. There is nothing in nature by which to conceive it or express it. All those operations in nature by which the presence of heat becomes known to us, cease before that point is quite reached. Hydrogen, the most difficult of gases to liquefy, is known to take the liquid form by measurement and calculation at 400° below, and the solid form at -418°. Beyond the temperature of the absolute zero, as now assumed, we have no ideas, out of all the universe, with which to conceive greater cold, since all the differences in nature and in matter, caused by different degrees of heat, have disappeared. Consequently, the absolute zero has been conceived and adopted as the starting point, a point below which the absence of heat could not fall at any point in the universe known to man.

Professor Dewar, who has carried working temperatures 150° farther downward than anybody before him, is still 115° distant from the great and ultimate end of his pilgrimage. He is gazing across the waste of untraversed cold in front of him, exactly like certain explorers in the Arctic Circle at the present time. He has gone as far as he can with the appliances which he has been able to devise, and knows it. To devise other appliances, or at least to find the key to other appliances which others may, perhaps, perfect, is his study. Oxygen has yet to be solidified. Hydrogen has still to be liquefied, and once hydrogen is liquefied the end will be near, as only 43° will then remain for conquest, and the means will probably then be at hand. In the meantime, Professor Dewar has been engaged for many months past in charting the strange country he has reached, building up stores, so to speak, of scientific knowledge, and aggregating facts for the use of the scientific world in general and the voyagers who will come after him. These are of many kinds, the dull routine which devolves upon every original investigator imposing hard and painstaking duties upon the scientific spirit very much like those of religion in their way. His first field of investigation, when his liquid oxygen gave him a working temperature of 280° below zero at will, was the alteration at lower and lower temperatures of the power of metals to conduct electricity. He found, as stated, that all converged toward a common point in this respect, and that at the temperature named they became nearly equal. Thus copper, which is an exceptionally good conductor at ordinary temperatures, and nickel, which is exceptionally bad, come closely together long before the absolute zero is approached. Alloys of metals, on the contrary, show a different and more eccentric action.

In strange contradistinction to these, carbon becomes a worse and worse conductor as the temperature is lowered, while at the temperature of the electric arc, the highest heat obtainable, its specific resistance becomes zero, and its power to conduct electricity reaches the maximum.

These experiments led to the belief that the molecular force or cohesion of metals also increased with lower and lower temperatures, and a large number of experiments have been made in this connection. The means adopted were special and ingenious. Professor Dewar found that the appliance best suited to his purpose was a cement-testing machine, the one used consisting principally of a lever, slowly and gently operated by weighting the long end with water added very gradually. The short end, as it rises, pulls apart a rod of the metal being tested. One of these machines is visible, in the accompanying picture, behind the lecture table. To test the breaking strength of little rods of various metals was not difficult. It was only necessary to put them in the clamps of the tester, and let the weight of water pull up the short end of the lever until the break occurred, and then jot down the reading on the index. When it came, however, to keeping these metals at a temperature of 280° below zero during all the time they were being tested, some ingenuity was required; in fact, a great deal of ingenuity. The experiment of testing a rod of frozen mercury, for instance, is very peculiar, and as it cannot be seen outside of Professor Dewar's laboratory, and gives a very good idea of the conditions therein, it may be of general interest. This large laboratory, now transformed into a workshop in cold, is always full, in working hours, of snowstorms. Every time liquid oxygen is drawn off from the condenser, a profuse cloud of snow fumes appears spreading in all directions. Nine-tenths of the oxygen, under its heavy pressure, vaporizes and scatters instantly upon reaching the air, and every molecule of it is so cold that snow, from the freezing of the moisture in the atmosphere, follows and marks its flying course. In the same way every open oxygen vessel has its own little snowstorm in unceasing progress, the white banks rising higher and higher about the necks of the jars, and settling upon any instruments or objects which have been cooled. A handy and common aid in the investigations is frozen carbonic acid. This was once a curiosity in itself. In the days—and they are not so very long ago—when 112° below zero was considered a marvel, frozen carbonic acid occupied the proud position of solid air at the present time. Now, however, the carbonic acid snow is simply put about in old bound-boxes, ready to be grabbed by the handful when needed and thrust about any metal which is in danger of getting warm. Human nature is so constituted that it can become accustomed in half an hour to railway travel at the rate of a mile a minute, and a man with a stop-watch on a Northern express is only interested when he is going at sixty-seven, seventy, or seventy-five miles an hour. In the same way, in this laboratory, frozen carbonic acid very quickly becomes commonplace, though it excites a momentary sympathy as of a material Napoleon in its Elba or a Paris exposition that has been outdone. To test the mercury, a small quantity of it is frozen in the oxygen until it becomes pasty. This is covered with frozen carbonic acid in a dish,

Professor Dewars's lecture table

and deftly moulded with little wooden spatulæ into the form of a little dumb-bell, the large ends being in this shape more easily gripped and held fast by the testing clamps. Fingers are not used, or they would be badly burned by the cold substances. Then it is plunged in the oxygen again and frozen hard. The whole of the clamping portion of the tester is then surrounded by the liquid oxygen and cooled to -280°. Then by delicate manipulation the rod of mercury is gripped, the breaking strain is applied, and the rod is broken, the tensile reading being, of course, a very low one as compared with more tenacious metals. The recent tests of the breaking strains of metals at -280° have given the following figures, which are advanced, not as mean averages, but as special experimental results from pure metals under the conditions of the time:

BREAKING STRESS.

TONS PER SQUARE INCH.

	59° Fahr.	-279.4° Fahr.
Copper	22.3	30
Iron	34	62.7
Brass	25.1	31.4
German silver	38.3	47
Steel	35.4	60

This increase in cohesive force at low temperatures has an interesting bearing astronomically. Cosmic dust and other theories of world formation which have at times been advanced, have been based to some extent upon the idea that when a planet died it underwent slow disintegration. Such an idea is entirely untenable, however, in view of the results of Professor Dewar's experiments upon the action of matter as it grows colder. The increase of cohesive force which he has made evident shows that the colder a planet grows, the more tenaciously its particles must be held together, and the harder and more enduring it must become. Nature, in fact, appears to embalm her dead planetary bodies throughout the universe very much more perfectly than the old Egyptians; so perfectly, in fact, that the great tomb of space must and will be filled with them forever, perfectly preserved unless some vagrant comet or eccentric star chances to knock them to pieces. If this earth, by any chance of the future, falls in temperature to the absolute zero, the solidified oxygen, nitrogen, carbonic acid gas, and water vapor of its atmosphere will cover it, over all its surface, with a white mantle from thirty-three to thirty-six feet thick.

The rays of the sun, however, as long as they shine upon it, will prevent the assumption of this frozen envelope, the estimated effect of sun rays upon the moon's temperature being, according to Newcome, a variation of 500°, from 200° below the Fahrenheit zero to 300° above, according to the periods and points of the moon's surface upon which the sun's heat acts.

Solid nitrogen is a white crystalline substance, and in both its liquid and solid states shows no variation from the inert properties which it displays in nature. Solid air is a transparent glass in which the nitrogen is solid and the liquid oxygen is held mechanically. It is obtained in the test tubes illustrated here with, each tube consisting of three compartments. The outer is the vacuum chamber. The second contains liquid oxygen, which is boiled off rapidly by exhausting the air. This produces such intense cold in the third tube, which is inserted in the liquid oxygen and is open to the air, that the air from without first liquefies and runs rapidly down the sides of the innermost tube, forming a clear liquid like water at the bottom. As the cold is increased the liquid grows thicker and thicker, and in a few moments is a solid, looking exactly like ice or glass. When the tube containing it is taken out, it liquefies and vaporizes rapidly, and more air from the atmosphere liquefies on the outside of the tube and drops freely from the bottom, passing into vapor, however, before it reaches the floor. Oxygen has not yet been reduced to a solid, and this achievement will be the next step downward. The present belief is that it will not solidify in crystalline form. The solidification of hydrogen, whenever in the future it takes place, will throw light on one of the most interesting problems in chemistry. Hydrogen, which we only know as the lightest and most elusive of all gases, is believed, and for good reasons, to be actually the vapor of a metal, and a true metal in all respects. Between its boiling point, 400° below zero Fahrenheit, and that of nearly all the other metals, which range from 1,200° to 3,500° above zero, the difference is wide and suggests the need of the absolute thermometer. As a gas, it forms alloys with metals precisely as if it were a metal, and under condensation it shows an increased power of conducting heat and electricity, the increase accurately following the metallic instead of the gaseous law. Faraday was the first to advance the theory that hydrogen was a metal, and he prophesied that, if it were ever solidified, it would have the texture and lustre of a metal, which view has not been opposed by any facts established since his time. Nitrous oxide and ethylene have given Professor Dewar liquid oxygen, and liquid oxygen has enabled him to reach solid air and -346°, the lowest temperature yet known. Beyond this he cannot go till invention or discovery open a way that is for the present impassable.

The molecular force called heat has its own special mysteries, however, and to illustrate this fact let the reader imagine himself for a time in the laboratory of Professor Dewar, viewing one of the simplest yet strangest examples of the work that is now being carried on. The amount of heat that is required to produce great cold is the first paradox that strikes him. Instead of ice and freezing mixtures, he is confronted with two steam-engines, one gas-engine, four big steel compressors, two large and powerful air-pumps, and all the wheels, shafts, and gearing which unite these powerful mechanical agencies. The whole idea is pressure. Compression forces certain gases to become liquids, and when the pressure is removed and they are permitted to return to the gaseous state, they absorb from their immediate surroundings as much heat, generally speaking, as was lost in liquefying them. Nitrous oxide gas is liquefied by a pressure of one thousand four hundred pounds to the square inch. This pressure was obtained by a steam-engine run by heat. When it gasifies again, it creates a cold of 130° below zero. The same amount of heat as was used up, in the form of obtained pressure, in liquefying it, it suddenly absorbs from its surroundings when it again becomes a gas. Within its chamber is a chamber of ethylene gas under a pressure of one thousand eight hundred pounds to the square inch, and the cold from the evaporation of the nitrous oxide liquefies the ethylene at this pressure. Within the ethylene chamber is the pipe from the oxygen compressor, the oxygen being compressed to seven hundred and fifty pounds to the square inch. When the ethylene gasifies or evaporates, it reduces the temperature to 229° below zero, and at this point the compressed oxygen liquefies freely and is drawn off in quantity at a cost—including the waste, which is inevitably about ninety per cent.—of perhaps five hundred dollars per gallon. This is only a rough estimate, to be sure, but there is no doubt, all things considered, that five hundred dollars per gallon as a selling price would have left no margin of profit if every quart of the beautiful pale blue liquid thus far obtained had been sold at this price.

This much is preliminary, and may not be new. Now, however, comes the simple but inconceivable strangeness of heat. A globular glass oxygen vessel, as round as a ball, is filled with the liquid. The narrow neck is open to the air, to prevent the explosion which would follow if it were confined, and which would be similar to the explosion of a steam-boiler with no valve. The oxygen simmers quietly, protected from the heat of the room by the glass compartment about it, an outer glass chamber, which is a perfect vacuum and conducts no heat. The liquid touches the air only at the neck, and the cold belt of extremely cold oxygen vapor at this point prevents any rapid evaporation. A powerful electric light is set in action beside the vessel, and the tremendous heat and light which it is giving off are sent through a lens, which bends the light and heat rays into parallel lines, and these are then sent through the oxygen. Their heat at the point of passage is great, far above the boiling point of water. The heat is so powerful that if it is focused by a burning-glass it will instantly set fire to wood or paper, and this heat is sent through liquid oxygen, a liquid so cold that it will freeze mercury, alcohol, or any known liquid— a cold of 280° below zero. It would certainly seem that so great a heat as this would instantly cause the liquid oxygen to become vapor, boil furiously, and burst the vessel. On the contrary, it has a very insignificant effect. The boiling shows some increase, but nothing like what would be expected. The strange thing is that the heat is *passing through* this great cold practically unaffected. It is coming out on the other side, because, as is quickly seen, the spherical glass vessel acts as a burning-glass and converges the rays, and a piece of paper placed at the focus bursts into flame. Heat, indeed, can pass through any degree of cold and still be heat. It has lost only twenty per cent. of its force in passing through this extremely cold medium, and only ten per cent. of the loss is due to the liquid oxygen, the other ten being absorbed by the glass. This illustrates how the sun's heat reaches the earth, passing through ninety-three millions of miles of cold so great as to be only imaginable, and still having the power when it reaches the earth to produce all the phenomena of life and natural force which are visible to us.

There are other experiments, curious and often beautiful. Soap-bubbles are frozen in the cold oxygen vapor, and rest, little iridescent globes of the filmiest ice, upon the gently simmering liquid below. Ice, our ordinary freezing medium, is often frozen in its turn, and, as a solid, shrinks, cracks, splits up, and goes to pieces at sudden cooling, exactly like hot glass in cold water. Liquid ozone, that eccentric, indigo-blue twin sister of oxygen, is being observed and studied in all its moods. And with Dewar working at 346° of cold, and Moissan, in Paris, investigating nature at 6,300° above the Fahrenheit zero, the world can scarcely fail before many years to gain new and perhaps startling knowledge of that comfortable, inconceivable, and inexplicable phenomenon, the force or motion or thing to which we have given the name of "heat."

Scientists

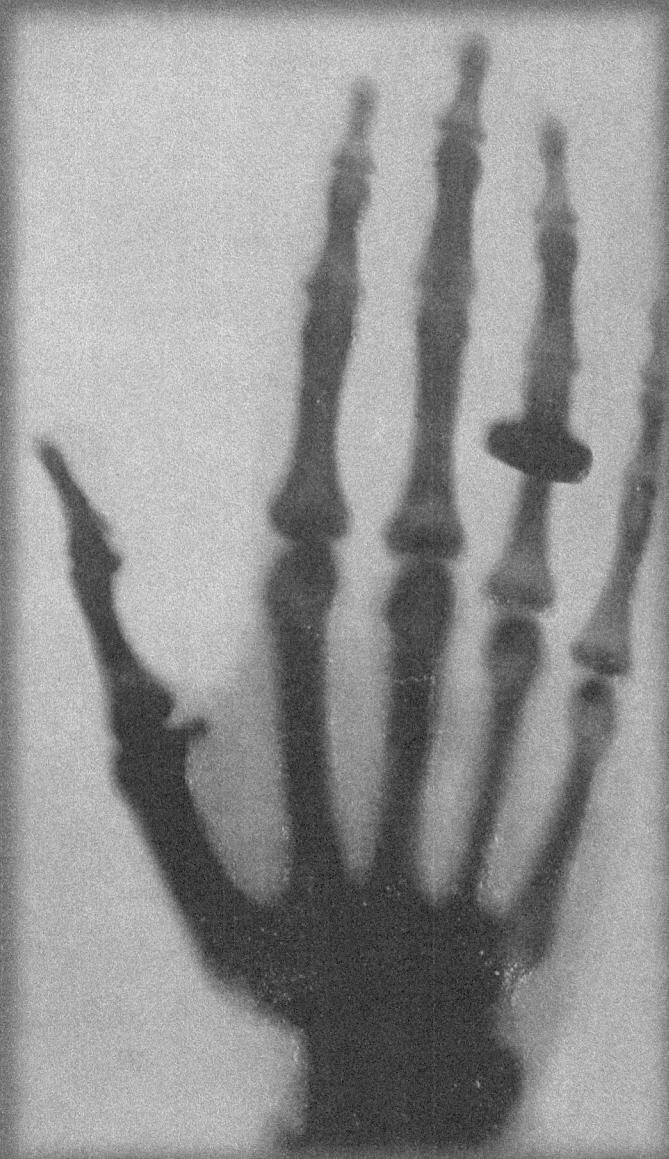

The Edge of the Future—An Interview with Professor Alexander Graham Bell

By Cleveland Moffett

June 1893

PROFESSOR Graham Bell is not like some pedantic wise men who talk as if they believed that the end of knowledge in their particular line had been already reached. On the contrary, this distinguished inventor is convinced that the discovery and inventions of the past will seem but trivial things when compared with those which are to come. Nor does he think that the day of man's greater knowledge is so very far distant.

THE AIR-SHIP OF THE NEAR FUTURE.

"I have not the shadow of a doubt"—these are his own words, spoken to me quite recently at Washington—"that the problem of aerial navigation will be solved within ten years. That means an entire revolution in the world's methods of transportation and of making war. I am able to speak with more authority on this subject from the fact of being actively associated with Professor Langley of the Smithsonian Institution in his researches and experiments. I am not at liberty to speak in detail of these experiments, but will say that the calculations of scientific men in regard to the amount of power necessary to maintain an air-ship above the earth have been strangely erroneous; I may say ridiculously so. According to these, Nature would have given the birds and insects a muscular force vastly greater and superior in its qualities to that bestowed upon man. That seems unreasonable in the first place, when one reflects that man is at the head of creation, and we have found practically that such is not the case. The power required to lift and propel an air-ship is very much less than has been supposed; indeed, Professor Langley concludes that when the air-ship has once been lifted above the earth to the proper height, it will be possible to maintain it there with proportionately no greater effort than that expended by hawks and eagles in sailing about with extended wings. The air strata will do the bulk of the lifting, if a small propelling power is provided. Of course, a greater power will be necessary to lift the air-ship originally, and it may be some time before the art of managing an air-ship is discovered; but the final result, I am convinced, will allow men to sail about in the air as easily and as safely as the birds do. I predict that we will see the beginning of this modern miracle by the end of the nineteenth century.

"Of course the air-ship of the future will be constructed without any balloon attachment. The discovery of the balloon undoubtedly retarded the solution of the flying problem for over a hundred years. Ever since the Montgolfiers taught the world how to rise in the air by means of inflated gas-bags, the inventors working at the problem of aerial navigation have been thrown on the wrong track. Scientific men have been wasting their time trying to steer balloons, a thing which in the nature of the case is impossible to any great extent, inasmuch as balloons, being lighter than the resisting air, can never make headway against it. The fundamental principle of aerial navigation is that the air-ship must be heavier than the air. It is only of recent years that men capable of studying the problem seriously have accepted this as an axiom. Electricity in one form or another will undoubtedly be the motive

Alexander Graham Bell. ca. 1902.
Courtesy Library of Congress.

power for air-ships, and every advance in electrical knowledge brings us one step nearer to the day when we shall fly. It would be perfectly possible, to-day, to direct a flying machine by means of pendant electric wires which would transmit the necessary current without increasing the load to be borne. Perhaps a feasible means of propelling such an air-ship would be by a kind of trolley system where the rod would hang down from the car to the stretched wire, instead of extending upward. This is an idea which I would recommend to inventors."

It is most interesting to watch Professor Bell as he talks about the great inventions which he sees with prophetic eye in store for the world. He has the happy faculty of expressing great ideas in simple words, and there is nothing ponderous in his speech. He is as enthusiastic as a school-boy thinking of the kite he will make as big as a barndoor. His black eyes flash, and they seem all the blacker contrasted with his white hair; the words tumble out quickly, and those who have the good fortune to listen are carried away by the magnetism of this great inventor.

SEEING BY ELECTRICITY.

The mention of electricity brought up new possibilities for future discovery, some of them so amazing as to almost pass the bounds of credibility. He said:

"Morse taught the world years ago to write at a distance by electricity; the telephone enables us to talk at a distance by electricity; and now scientists are agreed that there is no theoretical reason why the well-known principles of light should not be applied in the same way that the principles of sound have been applied in the telephone, and thus allow us to see at a distance by electricity. It is some ten years since the scientific papers of the world were greatly exercised over a report that I had filed at the Smithsonian Institution a sealed packet supposed to contain a method of doing this very thing; that is, transmit the vision of persons and things from one point on the earth to another. As a matter of fact, there was no truth in the report, but it resulted in stirring up a dozen scientific men of eminence to come out with statements to the effect that they too had discovered various methods of seeing by electricity. That shows what I know to be the case, that men are working at this great problem in many laboratories, and I firmly believe it will be solved one day.

"Of course, while the principle of seeing by electricity at a distance is precisely that applied in the telephone, yet it will be very much more difficult to construct such an apparatus, owing to the immensely greater rapidity with which the vibrations of light take place when compared with the vibrations of sound. It is merely a question, however, of finding a diaphragm which will be sufficiently sensitive to receive these vibrations and produce the corresponding electrical variations."

THOUGHT TRANSFERENCE BY ELECTRICITY.

After he had spoken of this idea for some time, Professor Bell stopped suddenly, and, with an amused twinkle in his eyes, exclaimed: "But while we are talking of all this, what is to prevent some one from discovering a way of thinking at a distance by electricity?"

Having said this, the genial professor threw himself back and laughed heartily at the amazement his words awakened. Was he joking? Apparently not, for he proceeded seriously to discuss one of the most astounding conceptions that ever entered an inventor's mind. Thinking by electricity! Imagine two persons, one thousand or ten thousand miles apart, placed in communication electrically, in such a way that, without any spoken word, without sounding-board, key, or any bodily movement, the one receives instantly the thoughts of the other, and instantly sends back his own thoughts. The wife in New York knows what is passing in the brain of her husband in Paris. The husband has the same knowledge. What boundless possibilities, to be sure, this arrangement offers for business men, lovers, humorous writers, and the police authorities!

Preposterous as such an idea appears in its first conception, it certainly assumes an increasing plausibility when one listens to Professor Bell's reasoning.

"After all," he says, "what would there be in such a system more mysterious than in the processes of the mind reader? You substitute a wire and batteries for a strange-eyed man in a dress suit, that is all."

The logical basis of Professor Bell's scheme is clear, and its details quite beautiful in their simplicity, when you admit his major premise. That premise is that the human brain is merely a kind of electrical reservoir, and that thinking is nothing more than an electrical disturbance, like the aurora borealis or the sparks from a Holtz machine. The nerves are the wires leading from the central battery in the head. The reasonableness of this assumption is increased when one remembers that electricity may be made to act upon the nerves, even in a lifeless body, so as to produce the same muscular contractions which are produced by the brain force, whatever that may be. We talk of animal magnetism. What if it were the same as any other kind of magnetism? If these two forces are identical in one respect, why may they not be so in all respects? So Professor Bell reasons, and granting that the human brain is merely a storehouse of electricity for our bodily needs, of electricity not essentially different from that which we know elsewhere, it must be possible to apply the same electrical laws to the brain as to any other electric apparatus and to get similar results.

"Do you begin to see my idea?" said Professor Bell, growing more and more enthusiastic as he proceeded. Then he gave a rapid outline of what might be a system of thinking by electricity.

Everyone knows, who knows anything about the subject, that an electric current passing inside of a coil of wire induces an electric current in that wire. Now, if the human brain be taken as a battery, then currents are constantly passing from it to various parts of the body, and the head may be considered in a state of constant electrical excitement, the intensity varying with the character of the thought processes. Now, suppose a coil of wire properly prepared in the shape of a helmet, and fitted about the head of one person, with wires attached and connected with a helmet similarly fitted upon the head of another person at any convenient distance. Every electric current in the one human battery must induce a current in the coil around the head, which current must be transmitted to the other coil. This other coil must then, by the reversed process, induce a current in the brain within helmet No. 2, and that person must receive some cerebral sensation. This cerebral sensation might be a thought, and probably would be, if it turns out to be true that brain force is identical with electricity. In that case, the thought of the one person would have produced a thought in the other person, and there is, if we go as far as this, every reason to believe that it would be the same thought. Thus the problem of thinking at a distance by electricity would be solved.

So much for a curious theory of what might be, if so and so were true; but Professor Bell has not stopped with theories, but has actually begun to put them to the test. Not that he is over-sanguine as to the result, but he believes the experiment worth the making, and that seriously. He has actually had two helmets, such as those described, constructed, and has begun a series of experiments in his laboratory. Thus far, the results have been for the most part negative, but not so much so as to prevent him hoping that more perfect appliances may lead to something more conclusive. It is true that the thought in one brain has produced a sensation in the other, through the two helmets, but what the relation was between the thought and the sensation could not be determined.

MAKING THE DEAF HEAR BY THE USE OF ELECTRICITY.

By quick stages the conversation ran into another channel with new wonders possible in the future. Professor Bell has conceived of a method of mak-ing the deaf hear, which is certainly startling. He proposes to do away with ears entirely, and produce the sensations of hearing by direct communication with the brain, through the bones of the head. As a matter of fact, the brains of deaf people are usually in a perfectly healthy condition, and the only thing which prevents them from hearing is some defect in communication with the vibrating air. If their brains could be excited artificially in the same way that the brains of ordinary persons are excited by vibrations communicated through the various chambers and passages of the ear, then the deaf would hear in the same way that other persons do.

It is, of course, a fact, that hearing in every instance is merely an illusion of the senses, a sort of tickling of the brain. This tickling of the brain is ordi-narily accomplished by the nerve force passing from the third chamber of the ear to the brain itself. If this nerve force is nothing more or less than ordinary electricity, and if science can train electricity to tickle the brain artificially in the same way and at the same points that the nerves from the ear usually do, then the ordinary sensations of hearing must result, whether the person has ears or not. The problem here is to discover the proper way of tickling the brain. The gentlemen who seat themselves in electrocution chairs have their brains tickled in a way which would not be generally satisfactory.

THERE IS DANGER IN SUCH EXPERIMENTS.

In his desire to bring relief to the deaf—and his whole life has been devot-ed to that object—Professor Bell has begun a series of remarkable experi-ments in this line. Some time ago, he determined to study the effects produced upon the brain by turning an electric current into it through the side of the head. With this end in view, he arranged a dynamo machine with a feeble cur-rent, giving a varying number of interruptions per second, and attached one of the poles to a wet sponge which he placed in one of his ears.

"I risked one of my ears," he said simply, "in making this experiment, but I could not risk them both, so I held the second pole of the machine in my hand and turned on the current."

Fortunately no harm resulted, but immediately Professor Bell experi-enced the sensation of a pleasant sound whose pitch he was able to vary by increasing or diminishing the number of interruptions in the dynamo ma-chine. His assistant standing beside him could detect no sound at all, so that what Professor Bell heard must have been the effect of the electric current upon his brain. This effect he found could be varied by varying the character of the current. Now he argues that greater variations might be produced in the sounds heard by the brain if the current turned into it were varied in the proper manner. For instance, suppose the current from a long distance telephone to be turned through the head of the deaf mute, a sponge connected with either pole being placed in each ear. Then let some one talk into the telephone in the ordinary way, the infinite variations in the current produced by the voice vibrations being passed into the brain directly. Is it not conceivable that such a variety of brain sensations or tones might then be caused in the head of the deaf mute as to make it possible to establish a system of sound signals, so to speak, which would be the equivalent of ordinary language? Indeed, is it not possible that the deaf mute might actually hear spoken words?

Professor Bell's experiments upon himself have been so encouraging as to make him disposed to try more complete experiments in the same line upon persons who have lost all sense of hearing, and who would doubtless be willing to take the inevitable risk for the sake of the great blessing which a successful issue would bring to them.

We talked a long time about these strange fancies, and finally I said to Professor Bell:

"But on this principle of brain tickling, what is to prevent a blind man from seeing by electricity?"

"I do not know that there is anything to prevent it."

The Edge of the Future—Unsolved Problems That Edison is Studying

By E. J. Edwards

June 1893

THOMAS A. EDISON, when he was congratulated upon his forty-sixth birthday, declared that he did not measure his life by years, but by achievements or by campaigns; and he then confessed that he had planned ahead many campaigns, and that he looks forward to no period of rest, believing that for him, at least, the happiest life is a life of work. In speaking of his campaigns Mr. Edison said: "I do not regard myself as a pure scientist, as so many persons have insisted that I am. I do not search for the laws of nature, and have made no great discoveries of such laws. I do not study science as Newton and Kepler and Faraday and Henry studied it, simply for the purpose of learning truth. I am only a professional inventor. My studies and experiments have been conducted entirely with the object of inventing that which will have commercial utility. I suppose I might be called a scientific inventor, as distinguished from a mechanical inventor, although really there is no distinction."

When Mr. Edison was asked about his campaigns and those achievements by which he measured his life, he said that in the past there had been first the stock-ticker and the telephone, upon the latter of which he worked very hard. But he regarded the greatest of his achievements, in the early part of his career, as the invention of the phonograph. "That," said he, "was an invention pure and simple. No suggestion of it, so far as I know, had ever been made; and it was a discovery made by accident, while experimenting upon another invention, that led to the development of the phonograph.

"My second campaign was that which resulted in the invention of the incandescent lamp. Of course, an incandescent lamp had been suggested before. There had been abortive attempts to make them, even before I knew anything

215

about telegraphing. The work which I did was to make an incandescent lamp which was commercially valuable, and the courts have recently sustained my claim to priority of invention of this lamp. I worked about three years upon that. Some of the experiments were very delicate and very difficult; some of them needed help which was very costly. That so far has been, I suppose, my chief achievement. It certainly was the first one which made me independent, and left me free to begin other campaigns without the necessity of calling for outside capital, or of finding my invention subjected to the mysteries of Wall Street manipulation."

The hint contained in Mr. Edison's reference to Wall Street, and the mysteries of financiering which prevail there, led naturally enough to a question as to Mr. Edison's future

Thomas A. Edison

Drawn expressly for McClure's Magazine by W.D. Stevens. September 30, 1897

purpose with regard to capitalists, and he said: "In my future campaigns I expect myself to control absolutely such inventions as I make. I am now fortunate enough to have capital of my own, and that I shall use in these campaigns. The most important of the campaigns I have in mind is one in which I have now been engaged for several years. I have long been satisfied that it was possible to invent an ore-concentrator which would vastly simplify the prevailing methods of extracting iron from earth and rock, and which would do it so much cheaper than those processes as to command the market. Of course I refer to magnetic iron ore. Some of the New Jersey mountains contain practically inexhaustible stores of this magnetic ore, but it has been expensive to mine. I was able to secure mining options upon nearly all these properties, and then I began the campaign of developing an ore-concentrator which would make these deposits profitably available. This iron is unlike any other iron ore. It takes four tons of the ore to produce one ton of pure iron, and yet I saw, some years ago, that if some method of extracting this ore could be devised, and the mines controlled, an enormously profitable business would be developed, and yet a cheaper iron ore—cheaper in its first cost—would be put upon the market. I worked very hard upon this problem, and in one sense successfully, for I have been able by my methods to extract this magnetic ore at comparatively small cost, and deliver from my mills pure iron bricklets. Yet I have not been satisfied with the methods; and some months ago I decided to abandon the old methods and to undertake to do this work by an entirely new system. I had some ten important details to master before I could get a perfect machine, and I have already mastered eight of them. Only two remain to be solved; and when this work is complete, I shall have, I think, a plant and mining privileges which will outrank the incandescent lamp as a commercial venture, certainly so far as I am myself concerned. Whatever the profits are, I shall myself control them, as I have taken no capitalists in with me in this scheme."

Mr. Edison was asked if he was willing to be more explicit respecting this invention, but he declined to be, further than to say: "When the machinery is done as I expect to develop it, it will be capable of handling twenty thousand tons of ore a day with two shifts of men, five in a shift. That is to say, ten workmen, working twenty hours a day in the aggregate, will be able to take this ore, crush it, reduce the iron to cement-like proportions, extract it from the rock and earth, and make it into bricklets of pure iron, and do it so cheaply that it will command the market for magnetic iron."

Mr. Edison, in speaking of this campaign, referred to it as though it was practically finished; and it was evident in the conversation that already his mind turns to a new campaign, which he will take up as soon as his iron-ore concentrator is complete and its work can be left to competent subordinates.

He was asked if he would be willing to say what he had in mind for the next campaign, and he replied: "Well, I think as soon as the ore concentrating business is developed and can take care of itself, I shall turn my attention to one of the greatest problems that I have ever thought of solving, and that is, the direct control of the energy which is stored up in coal, so that it may be employed without waste and at a very small margin of cost. Ninety per cent of the energy that exists in coal is now lost in converting it into power. It goes off in heat through the chimneys of boiler-rooms. You perceive it when you step into a room where there is a furnace and boiler; it is also greatly wasted in the development of the latent heat which is created by the change from water to steam. Now that is an awful waste, and even a child can see that if this wastage can be saved, it will result in vastly cheapening the cost of everything which is manufactured by electric or steam power. In fact, it will vastly cheapen the cost of all the necessaries and luxuries of life, and I suppose the results would be of mightier influence upon civilization than the development of the steam-engine and electricity have been. It will, in fact, do away with steam-engines and boilers, and make the use of steam power as much of a tradition as the stage-coach now is.

"It would enable an ocean steamship of twenty thousand horse-power to cross the ocean faster than any of the crack vessels now do, and require the burning of only two hundred and fifty tons of coal instead of three thousand, which are now required; so that, of course, the charges for freight and passenger fares would be greatly reduced. It would enormously lessen the cost of manufacturing and of traffic. It would develop the electric current directly from coal, so that the cost of steam-engines and boilers would be eliminated. I have thought of this problem very much, and I have already my theory of the experiments, or some of them, which may be necessary to develop this direct use of all the power that is stored in coal. I can only say now, that the coal would be put into a receptacle, the agencies then applied which would develop its energy and save it all, and through this energy electric power of any degree desired could be furnished. Yes, it can be done; I am sure of that. Some of the details I have already mastered, I think; at least, I am sure that I know the way to go to work to master them. I believe that I shall make this my next campaign. It may be years before it is finished, and it may not be a very long time."

Mr. Edison looks farther ahead than this campaign, for he said: "I think it quite likely that I may try to develop a plan for marine signalling. I have the idea already pretty well formulated in my mind. I should use the well-known principle that water is a more perfect medium for carrying vibrations than air, and should develop instruments which may be carried upon sea-going vessels, by which they can transmit or receive, through an international code of signals, reports within a radius of say ten miles."

Mr. Edison believes that Chicago is to become the London of America early in the next century, while New York will be its Liverpool, and he is of opinion that very likely a ship canal may connect Chicago with tide water, so that it will itself become a great seaport.

There is a common impression that Mr. Edison is an agnostic, but he denies it; and he said, in closing the conversation, "I tell you that no person can be brought into close contact with the mysteries of nature, or make a study of chemistry, without being convinced that behind it all there is supreme intelligence. I am convinced of that, and I think that I could, perhaps

I may sometime, demonstrate the existence of such intelligence through the operation of these mysterious laws with the certainty of a demonstration in mathematics."

The New Marvel in Photography

A Visit to Professor Röntgen at His Laboratory In Würzburg.—His Own Account of His Great Discovery.—Interesting Experiments wth the Cathode Rays.—Practical Uses of the New Photography.

By H. J. W. Dam.

April 1896

IN all the history of scientific discovery there has never been, perhaps, so general, rapid, and dramatic an effect wrought on the scientific centres of Europe as has followed, in the past four weeks, upon an announcement made to the Würzburg Physico-Medical Society, at their December meeting, by Professor William Konrad Röntgen, professor of physics at the Royal University of Würzburg. The first news which reached London was by telegraph from Vienna to the effect that a Professor Röntgen, until then the possessor of only a local fame in the town mentioned, had discovered a new kind of light, which penetrated and photographed through everything. This news was received with a mild interest, some amusement, and much incredulity; and a week passed. Then, by mail and telegraph, came daily clear indications of the stir which the discovery was making in all the great line of universities between Vienna and Berlin. Then Röntgen's own report arrived, so cool, so business-like, and so truly scientific in character, that it left no doubt either of the truth or of the great importance of the preceding reports. To-day, four weeks after the announcement, Röntgen's name is apparently in every scientific publication issued this week in Europe; and accounts of his experiments, of the experiments of others following his method, and of theories as to the strange new force which he has been the first to observe, fill pages of every scientific journal that comes to hand. And before the necessary time elapses for this article to attain publication in America, it is in all ways probable that the laboratories and lecture-rooms of the United States will also be giving full evidence of this contagious arousal of interest over a discovery so strange that its importance cannot yet be measured, its utility be even prophesied, or its ultimate effect upon long-established scientific beliefs be even vaguely foretold.

The Röntgen rays are certain invisible rays resembling, in many respects, rays of light, which are set free when a high pressure electric current is discharged through a vacuum tube. A vacuum tube is a glass tube from which all the air, down to one-millionth of an atmosphere, has been exhausted after the insertion of a platinum wire in either end of the tube for connection with the two poles of a battery or induction coil. When the discharge is sent through the tube, there proceeds from the an-

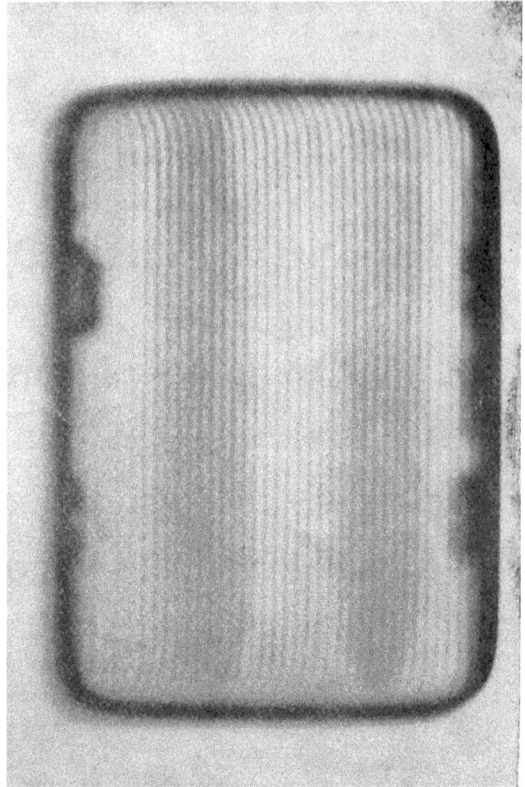

Picture of an aluminium cigar-case, showing cigars within.

From a photograph by A. A. C. Swinton, Victoria Street, London. Exposure, ten minutes.

ode—that is, the wire which is connected with the positive pole of the battery—certain bands of light, varying in color with the color of the glass. But these are insignificant in comparison with the brilliant glow which shoots from the cathode, or negative wire. This glow excites brilliant phosphorescence in glass and many substances, and these "cathode rays," as they are called, were observed and studied by Hertz; and more deeply by his assistant, Professor Lenard, Lenard having, in 1894, reported that the cathode rays would penetrate thin films of aluminium, wood, and other substances, and produce photographic results beyond. It was left, however, for Professor Röntgen to discover that during the discharge another kind of rays are set free, which differ greatly from those described by Lenard as cathode rays The most marked difference between the two is the fact that Röntgen rays are not deflected by a magnet, indicating a very essential difference, while their range

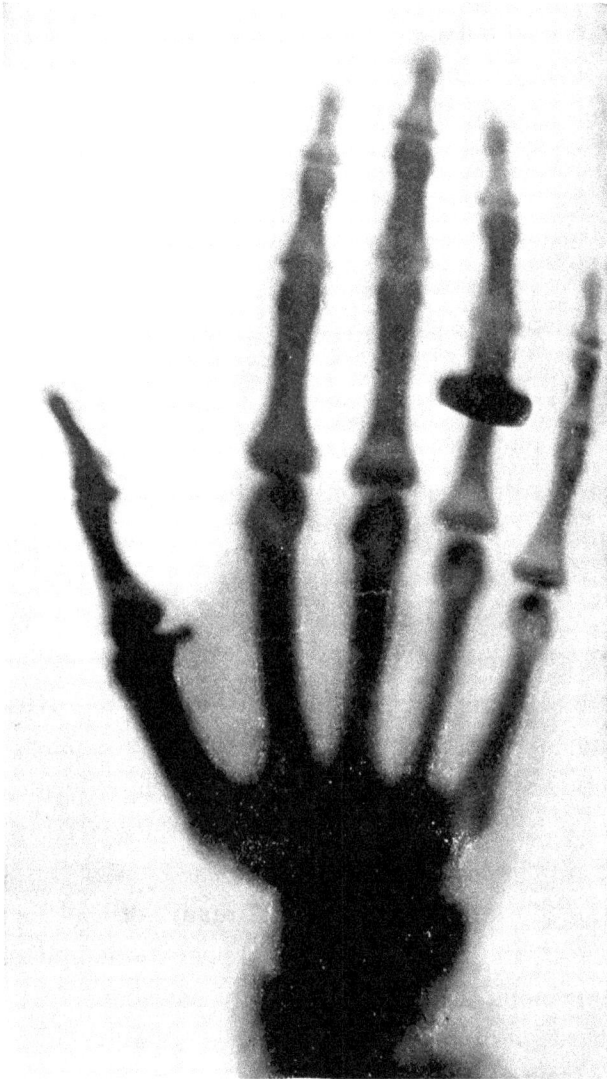

Photograph of a lady's hand, showing the bones , and a ring on the third finger with faint outlines of the flesh.

From a photograph by Mr. P. Spies, director of the "Urania," Berlin.

and penetrative power are incomparably greater. In fact, all those qualities which have lent a sensational character to the discovery of Röntgen's rays were mainly absent from these of Lenard, to the end that, although Röntgen has not been working in an entirely new field, he has by common accord been freely granted all the honors of a great discovery.

Exactly what kind of a force Professor Röntgen has discovered he does not know. As will be seen below, he declines to call it a new kind of light, or a new form of electricity. He has given it the name of the X rays. Others speak of it as the Röntgen rays. Thus far its results only, and not its essence, are known. In the terminology of science it is generally called "a new mode of motion," or, in other words, a new force. As to whether it is or not actually a force new to science, or one of the known forces masquerading under strange conditions, weighty authorities are already arguing. More than one eminent scientist has already affected to see in it a key to the great mystery of the law of gravity. All who have expressed themselves in print have admitted, with more or less frankness, that, in view of Röntgen's discovery, science must forthwith revise, possibly to a revolutionary degree, the long accepted theories concerning the phenomena of light and sound. That the X rays, in their mode of action, combine a strange resemblance to both sound and light vibrations, and are destined to materially affect, if they do not greatly alter, our views of both phenomena, is already certain; and beyond this is the opening into a new and unknown field of physical knowledge, concerning which speculation is already eager, and experimental investigation already in hand, in London, Paris, Berlin, and, perhaps, to a greater or less extent, in every well-equipped physical laboratory in Europe.

This is the present scientific aspect of the discovery. But, unlike most epoch-making results from laboratories, this discovery is one which, to a very unusual degree, is within the grasp of the popular and non-technical imagination. Among the other kinds of matter which these rays penetrate with ease is the human flesh. That a new photography has suddenly arisen which can photograph the bones, and, before long, the organs of the human body; that a light has been found which can penetrate, so as to make a photographic record, through everything from a purse or a pocket to the walls of a room or a house, is news which cannot fail to startle everybody. That the eye of the physician or surgeon, long baffled by the skin, and vainly seeking to penetrate the unfortunate darkness of the human body, is now to be supplemented by a camera, making all the parts of the human body as visible, in a way, as the exterior, appears certainly to be a greater blessing to humanity than even the Listerian antiseptic system of surgery; and its benefits must inevitably be greater than those conferred by Lister, great as the latter have been. Already, in the few weeks since Röntgen's announcement, the results of surgical operations under the new system are growing voluminous. In Berlin, not only new bone fractures are being immediately photographed, but joined fractures, as well, in order to examine the results of recent surgical work. In Vienna, imbedded bullets are being photographed, instead of being probed for, and extracted with comparative ease. In London, a wounded sailor, completely paralyzed, whose injury was a mystery, has been saved by the photographing of an object imbedded in the spine, which, upon extraction, proved to be a small knife-blade. Operations for malformations, hitherto obscure, but now clearly revealed by the new photography, are already becoming common, and are being reported from all directions. Professor Czermark of Graz has photographed the living skull, denuded of flesh and hair, and has begun the adaptation of the new photography to brain study. The relation of the new rays to thought rays is being eagerly discussed in what may be called the non-exact circles and journals; and all that numerous group of inquirers into the occult, the believers in clairvoyance, spiritualism, telepathy, and kindred orders of alleged phenomena, are confident of finding in the new force long-sought facts in proof of their claims. Professor Neusser in Vienna has photographed gall-stones in the liver of one patient (the stone showing snow-white in the negative), and a stone in the bladder of another patient. His results so far induce him to announce that all the organs of the human body can, and will, shortly, be photographed. Lannelongue of Paris has exhibited to the Academy of Science photographs of bones showing inherited tuberculosis which had not otherwise revealed itself. Berlin has already formed a society of forty for the immediate prosecution of researches into both "the character of the new force and its physiological possibilities. In the next few weeks these strange announcements will be trebled or quadrupled, giving the best evidence from all quarters of the great future that awaits the Röntgen rays,

Dr William Konrad Röntgen , discoverer of the X-rays.

From a photograph by Hanfstaengl, Frankfort-on-the-Main.

and the startling impetus to the universal search for knowledge that has come at the close of the nineteenth century from the modest little laboratory in the Pleicher Ring at Würzburg.

On instruction by cable from the editor of this magazine, on the first announcement of the discovery, I set out for Würzburg to see the discoverer and his laboratory. I found a neat and thriving Bavarian city of forty-five thousand inhabitants, which, for some ten centuries, has made no salient claim upon the admiration of the world, except for the elaborateness of its mediæval castle and the excellence of its local beer. Its streets were adorned with large numbers of students, all wearing either scarlet, green, or blue caps, and an extremely serious expression, suggesting much intensity either in the contemplation of Röntgen rays or of the beer aforesaid. All knew the residence of Professor Röntgen (pronunciation: "Renken"), and directed me to the "Pleicher Ring." The various buildings of the university are scattered in different parts

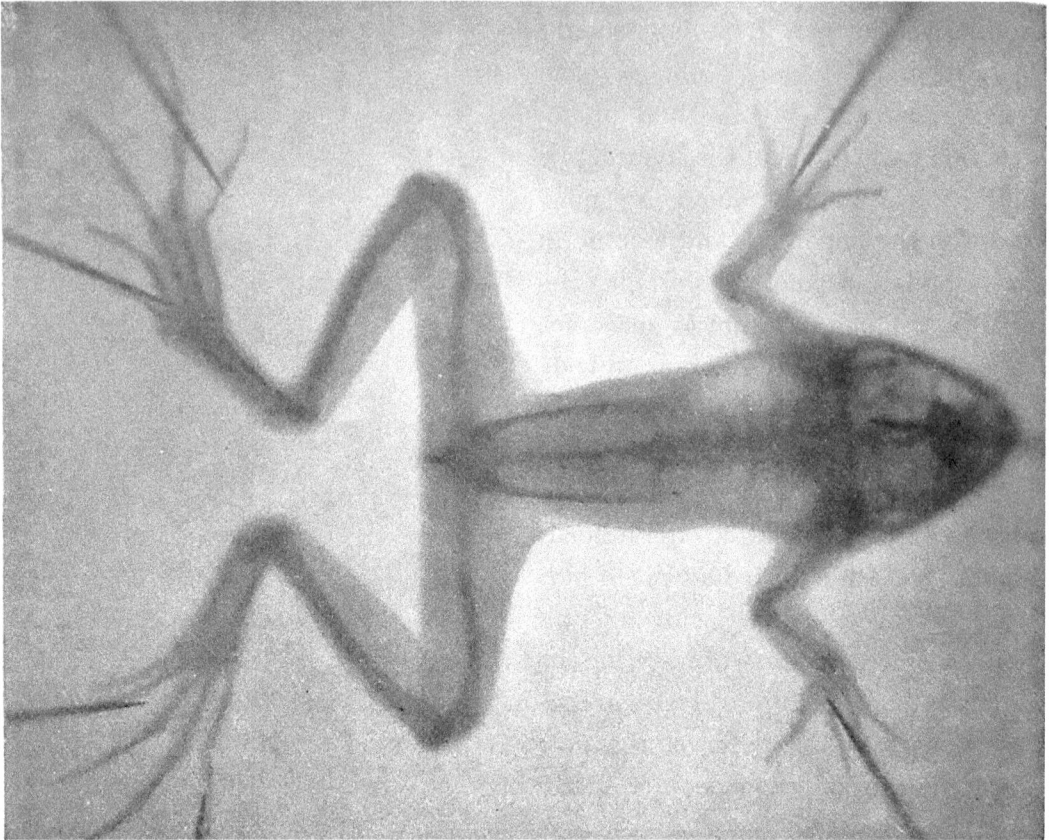

Skeleton of a frog, photographed through the flesh. The shadings indicate, in addition to bones, also the lungs and the cerebral lobes.

From a photograph by Professors Imbert and Bertin-Sans; reproduced by the courtesy of the "Presse Medicale," Paris.

In taking this photograph the experiment was tried of using a diaphragm interposed between the Crookes tube and the plate; and the superior clearness obtained is thought to result from this.

of Würzburg, the majority being in the Pleicher Ring, which is a fine avenue, with a park along one side of it, in the centre of the town. The Physical Institute, Professor Röntgen's particular domain, is a modest building of two stories and basement, the upper story constituting his private residence, and the remainder of the building being given over to lecture rooms, laboratories, and their attendant offices. At the door I was met by an old serving-man of the idolatrous order, whose pain was apparent when I asked for "Professor" Röntgen, and he gently corrected me with "Herr Doctor Röntgen." As it was evident, however, that we referred to the same person, he conducted me along a wide, bare hall, running the length of the building, with blackboards and charts on the walls. At the end he showed me into a small room on the right. This contained a large table desk, and a small table by the window, covered with photographs, while the walls held rows of shelves laden with laboratory and other records. An open door led into a somewhat larger room, perhaps twenty feet by fifteen, and I found myself gazing into a laboratory which was the scene of the discovery—a laboratory which, though in all ways modest, is destined to be enduringly historical.

There was a wide table shelf running along the farther side, in front of the two windows, which were high, and gave plenty of light. In the centre was a stove; on the left, a small cabinet, whose shelves held the small objects which the professor had been using. There was a table in the left-hand corner; and another small table—the one on which living bones were first photographed—was near the stove, and a Rhumkorff coil was on the right. The lesson of the laboratory was eloquent. Compared, for instance, with the elaborate, expensive, and complete apparatus of, say, the University of London, or of any of the great American universities, it was bare and unassuming to a degree. It mutely said that in the great march of science it is the genius of man, and not the perfection of appliances, that breaks new ground in the great territory of the unknown. It also caused one to wonder at and endeavor to imagine the great things which are to be done through elaborate appliances with the Röntgen rays—a field in which the United States, with its foremost genius in invention, will very possibly, if not probably, take the lead—when the discoverer himself had done so much with so little. Already, in a few weeks, a skilled London operator, Mr. A. A. C. Swinton, has reduced the necessary time of exposure for Röntgen photographs from fifteen minutes to four. He used, however, a Tesla oil coil, discharged by twelve half-gallon Leyden jars, with an alternating current of twenty thousand volts' pressure. Here were no oil coils, Leyden jars, or specially elaborate and expensive machines. There were only a Rhumkorff coil and Crookes (vacuum) tube and the man himself.

Professor Röntgen entered hurriedly, something like an amiable gust of wind. He is a tall, slender, and loose-limbed man, whose whole appearance bespeaks enthusiasm and energy. He wore a dark blue sack suit, and his long, dark hair stood straight up from his forehead, as if he were permanently electrified by his own enthusiasm. His voice is full and deep, he speaks rapidly, and, altogether, he seems clearly a man who, once upon the track of a mystery which appealed to him, would pursue it with unremitting vigor. His eyes are kind, quick, and penetrating; and there is no doubt that he much prefers gazing at a Crookes tube to beholding a visitor, visitors at present robbing him of much valued time. The meeting was by appointment, however, and his greeting was cordial and hearty. In addition to his own language he speaks French well and English scientifically, which is different from speaking it popularly. These three tongues being more or less within the equipment of his visitor, the conversation proceeded on an international or polyglot basis, so to speak, varying at necessity's demand.

It transpired, in the course of inquiry, that the professor is a married man and fifty years of age, though his eyes have the enthusiasm of twenty-five. He was born near Zurich, and educated there, and completed his studies and took his degree at Utrecht. He has been at Würzburg about seven years, and had made no discoveries which he considered of great importance prior to the one under consideration. These details were given under good-natured protest, he failing to understand why his personality should interest the public. He declined to admire himself or his results in any degree, and laughed at the idea of being famous. The professor is too deeply interested in science to waste any time in thinking about himself. His emperor had *fêted*, flattered, and decorated him, and he was loyally grateful. It was evident, however, that fame and applause had small attractions for him, compared to the mysteries still hidden in the vacuum tubes of the other room.

"Now, then," said he, smiling, and with some impatience, when the preliminary questions at which he chafed were over, "you have come to see the invisible rays."

"Is the invisible visible?"

"Not to the eye; but its results are. Come in here."

He led the way to the other square room mentioned, and indicated the induction coil with which his researches were made, an ordinary Rhumkorff coil, with a spark of from four to six inches, charged by a current of twenty amperes. Two wires led from the coil, through an open door, into a smaller room on the right. In this room was a small table carrying a Crookes tube connected with the coil. The most striking object in the room, however, was a huge and mysterious tin box about seven feet high and four feet square. It stood on end, like a huge packing-case, its side being perhaps five inches from the Crookes tube.

The professor explained the mystery of the tin box, to the effect that it was a device of his own for obtaining a portable dark-room. When he began his investigations he used the whole room, as was shown by the heavy blinds and curtains so arranged as to exclude the entrance of all interfering light from the windows. In the side of the tin box, at the point immediately against the tube, was a circular sheet of aluminium one millimetre in thickness, and perhaps eighteen inches in diameter, soldered to the surrounding tin. To study his rays the professor had only to turn on the current, enter the box, close the door, and in perfect darkness inspect only such light or light effects as he had a right to consider his own, hiding his light, in fact, not under the Biblical bushel, but in a more commodious box.

"Step inside," said he, opening the door, which was on the side of the box farthest from the tube. I immediately did so, not altogether certain whether my skeleton was to be photographed for general inspection, or my secret thoughts held up to light on a glass plate. "You will find a sheet of barium paper on the shelf," he added, and then went away to the coil. The door was closed, and the interior of the box became black darkness. The first thing I found was a wooden stool, on which I resolved to sit. Then I found the shelf on the side next the tube, and then the sheet of paper prepared with barium platino-cyanide. I was thus being shown the first phenomenon which attracted the discoverer's attention and led to the discovery, namely, the passage of rays, themselves wholly invisible, whose presence was only indicated by the effect they produced on a piece of sensitized photographic paper.

A moment later, the black darkness was penetrated by the rapid snapping sound of the high-pressure current in action, and I knew that the tube outside was glowing. I held the sheet vertically on the shelf, perhaps four inches from the plate. There was no change, however, and nothing was visible.

"Do you see anything?" he called.

"No."

"The tension is not high enough;" and he proceeded to increase the pressure by operating an apparatus of mercury in long vertical tubes acted upon automatically by a weight lever which stood near the coil. In a few moments the sound of the discharge again began, and then I made my first acquaintance with the Röntgen rays.

The moment the current passed, the paper began to glow. A yellowish-green light spread all over its surface in clouds, waves, and flashes. The yellow-green luminescence, all the stranger and stronger in the darkness, trembled, wavered, and floated over the paper, in rhythm with the snapping of the discharge. Through the metal plate, the paper, myself, and the tin box, the invisible rays were flying, with an effect strange, interesting, and uncanny. The metal plate seemed to offer no appreciable resistance to the flying force, and the light was as rich and full as if nothing lay between the paper and the tube.

"Put the book up," said the professor.

I felt upon the shelf, in the darkness, a heavy book, two inches in thickness, and placed this against the plate. It made no difference. The rays flew through the metal and the book as if neither had been there, and the waves of light, rolling cloud-like over the paper, showed no change in brightness. It was a clear, material illustration of the ease with which paper and wood are penetrated. And then I laid book and paper down, and put my eyes against the rays. All was blackness, and I neither saw nor felt anything. The discharge was in full force, and the rays were flying through my head, and, for all I knew, through the side of the box behind me. But they were invisible and impalpable. They gave no sensation whatever. Whatever the mysterious rays may be, they are not to be seen, and are to be judged only by their works.

I was loath to leave this historical tin box, but time pressed. I thanked the professor, who was happy in the reality of his discovery and the music of his sparks. Then I said: "Where did you first photograph living bones?"

"Here," he said, leading the way into the room where the coil stood. He pointed to a table on which was another—the latter a small short-legged wooden one with more the shape and size of a wooden seat. It was two feet square and painted coal black. I viewed it with interest. I would have bought it, for the little table on which light was first sent through the human body will some day be a great historical curiosity; but it was "nicht zu verkaufen." A photograph of it would have been a consolation, but for several reasons one was not to be had at present. However, the historical table was there, and was duly inspected.

"How did you take the first hand photograph?" I asked.

The professor went over to a shelf by the window, where lay a number of prepared glass plates, closely wrapped in black paper. He put a Crookes tube underneath the table, a few inches from the under side of its top. Then he laid his hand flat on the top of the table, and placed the glass plate loosely on his hand.

"You ought to have your portrait painted in that attitude," I suggested.

"No, that is nonsense," said he, smiling.

"Or be photographed." This suggestion was made with a deeply hidden purpose.

The rays from the Röntgen eyes instantly penetrated the deeply hidden purpose. "Oh, no," said he; "I can't let you make pictures of me. I am too busy." Clearly the professor was entirely too modest to gratify the wishes of the curious world.

"Now, Professor," said I, "will you tell me the history of the discovery?"

"There is no history," he said. "I have been for a long time interested in the problem of the cathode rays from a vacuum tube as studied by Hertz and Lenard. I had followed theirs and other researches with great interest, and determined, as soon as I had the time, to make some researches of my own. This time I found at the close of last October. I had been at work for some days when I discovered something new."

"What was the date?"

"The eighth of November."

"And what was the discovery?"

"I was working with a Crookes tube covered by a shield of black card-board. A piece of barium platino-cyanide paper lay on the bench there. I had been passing a current through the tube, and I noticed a peculiar black line across the paper."

"What of that?"

"The effect was one which could only be produced, in ordinary parlance,

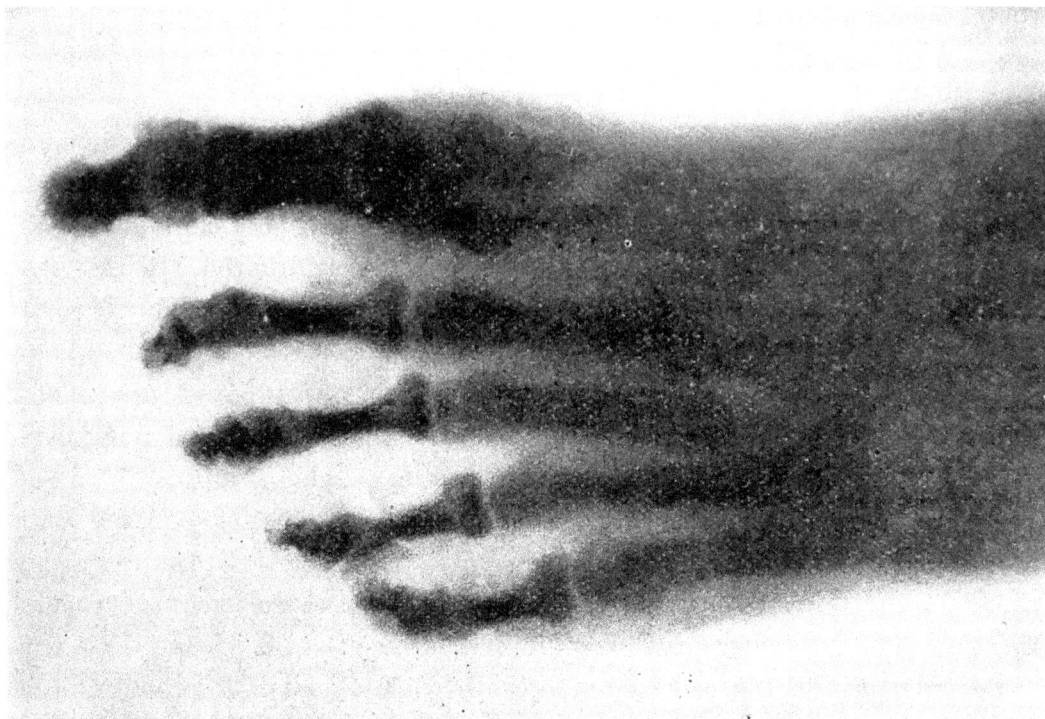

Bones of a human foot photographed through the flesh.

From a photograph by A. A. C. Swinton, Victoria Street, London. Exposure, fifty-five seconds.

by the passage of light. No light could come from the tube, because the shield which covered it was impervious to any light known, even that of the electric arc."

"And what did you think?"

"I did not think; I investigated. I assumed that the effect must have come from the tube, since its character indicated that it could come from nowhere else. I tested it. In a few minutes there was no doubt about it. Rays were coming from the tube which had a luminescent effect upon the paper. I tried it successfully at greater and greater distances, even at two metres. It seemed at first a new kind of invisible light. It was clearly something new, something unrecorded."

"Is it light?"

"No."

"Is it electricity?"

"Not in any known form."

"What is it?"

"I don't know."

And the discoverer of the X rays thus stated as calmly his ignorance of their essence as has everybody else who has written on the phenomena thus far.

"Having discovered the existence of a new kind of rays, I of course began to investigate what they would do." He took up a series of cabinet-sized photographs. "It soon appeared from tests that the rays had penetrative power to a degree hitherto unknown. They penetrated paper, wood, and cloth with ease; and the thickness of the substance made no perceptible difference, within reasonable limits." He showed photographs of a box of laboratory weights of platinum, aluminium, and brass, they and the brass hinges all having been photographed from a closed box, without any indication of the box. Also a photograph of a coil of fine wire, wound on a wooden spool, the wire having been photographed, and the wood omitted. "The rays," he continued, "passed through all the metals tested, with a facility varying, roughly speaking, with the density of the metal. These phenomena I have discussed carefully in my report to the Würzburg society, and you will find all the technical results therein stated." He showed a photograph of a small sheet of zinc. This was composed of smaller plates soldered laterally with solders of different metallic proportions. The differing lines of shadow, caused by the difference in the solders, were visible evidence that a new means of detecting flaws and chemical variations in metals had been found. A photograph of a compass showed the needle and dial taken through the closed brass cover. The markings of the dial were in red metallic paint, and thus interfered with the rays, and were reproduced. "Since the rays had this great penetrative power, it seemed natural that they should penetrate flesh, and so it proved in photographing the hand, as I showed you."

A detailed discussion of the characteristics of his rays the professor considered unprofitable and unnecessary. He believes, though, that these mysterious radiations are not light, because their behavior is essentially different from that of light rays, even those light rays which are themselves invisible. The Röntgen rays cannot be reflected by reflecting surfaces, concentrated by lenses, or refracted or diffracted. They produce photographic action on a sensitive film, but their action is weak as yet, and herein lies the first important field of their development. The professor's exposures were comparatively long—an average of fifteen minutes in easily penetrable media, and half an hour or more in photographing the bones of the hand. Concerning vacuum tubes, he said that he preferred the Hittorf, because it had the most perfect vacuum, the highest degree of air exhaustion being the consummation most desirable. In answer to a question, "What of the future?" he said:

"I am not a prophet, and I am opposed to prophesying. I am pursuing my investigations, and as fast as my results are verified I shall make them public."

"Do you think the rays can be so modified as to photograph the organs of the human body?"

In answer he took up the photograph of the box of weights. "Here are already modifications," he said, indicating the various degrees of shadow produced by the aluminium, platinum, and brass weights, the brass hinges, and even the metallic stamped lettering on the cover of the box, which was faintly perceptible.

"But Professor Neusser has already announced that the photographing of the various organs is possible."

"We shall see what we shall see," he said. "We have the start now; the developments will follow in time."

"You know the apparatus for introducing the electric light into the stomach?"

"Yes."

"Do you think that this electric light will become a vacuum tube for photographing, from the stomach, any part of the abdomen or thorax?"

The idea of swallowing a Crookes tube, and sending a high frequency current down into one's stomach, seemed to him exceedingly funny. "When I have done it, I will tell you, " he said, smiling, resolute in abiding by results.

"There is much to do, and I am busy, very busy, " he said in conclusion. He extended his hand in farewell, his eyes already wandering toward his work in the inside room. And his visitor promptly left him; the words, "I am busy," said in all sincerity, seeming to describe in a single phrase the essence of his character and the watchword of a very unusual man.

Returning by way of Berlin, I called upon Herr Spies of the Urania, whose photographs after the Röntgen method were the first made public, and have been the best seen thus far. The Urania is a peculiar institution, and one which it seems might be profitably duplicated in other countries. It is a scientific theatre. By means of the lantern and an admirable equipment of scientific appliances, all new discoveries, as well as ordinary interesting and picturesque phenomena, when new discoveries are lacking, are described and illustrated daily to the public, who pay for seats as in an ordinary theatre, and keep the Urania profitably filled all the year round. Professor Spies is a young man of great mental alertness and mechanical resource. It is the photograph of a hand, his wife's hand, which illustrates, perhaps better than any other illustration in this article, the clear delineation of the bones which can be obtained by the Röntgen rays. In speaking of the discovery he said:

"I applied it, as soon as the penetration of flesh was apparent, to the photograph of a man's hand. Something in it had pained him for years, and the photograph at once exhibited a small foreign object, as you can see;" and he exhibited a copy of the photograph in question. "The speck there is a small piece of glass, which was immediately extracted, and which, in all probability, would have otherwise remained in the man's hand to the end of his days." All of which indicates that the needle which has pursued its travels in so many persons, through so many years, will be suppressed by the camera.

"My next object is to photograph the bones of the entire leg," continued Herr Spies. "I anticipate no difficulty, though it requires some thought in manipulation."

It will be seen that the Röntgen rays and their marvellous practical possibilities are still in their infancy. The first successful modification of the action of the rays so that the varying densities of bodily organs will enable them to be photographed, will bring all such morbid growths as tumors and cancers into the photographic field, to say nothing of vital organs which may be abnormally developed or degenerate. How much this means to medical and surgical practice it requires little imagination to conceive. Diagnosis, long a painfully uncertain science, has received an unexpected and wonderful assistant; and how greatly the world will benefit thereby, how much pain will be saved, and how many lives saved, the future can only determine. In science a new door has been opened where none was known to exist, and a side-light on phenomena has appeared, of which the results may prove as penetrating and astonishing as the Röntgen rays themselves. The most agreeable feature of the discovery is the opportunity it gives for other hands to help; and the work of these hands will add many new words to the dictionaries, many new facts to science, and, in the years long ahead of us, fill many more volumes than there are paragraphs in this brief and imperfect account.

John Milne: Observer of Earthquakes

In an Earthquake Observatory.—Plotting the Greater Hollows of the Sea.

By Cleveland Moffett

May 1898

A T the very center of the Isle of Wight, in a little place called Shide, that most people in England never heard of, lives a scientist who probably knows more about earthquakes than any one else in the world—John Milne, member of learned societies, late professor of seismology at the University of Tokio, and a charming man into the bargain. His house looks down upon the roads where the Queen drives daily while at Osborne, and not far distant rise the towers of Carisbrooke Castle, where Charles I. was a prisoner.

Here, on a quiet hill, grown over with old trees and banks of ivy, away from all rush and noise, Professor Milne may be found, as I found him, work-ing among strange instruments of his own devising, operated by clockwork and electricity, and possessing such sensitiveness that an earthquake shock in Borneo will set them swinging for hours. With these wonderful pendulums, of which I shall speak presently, the Professor watches throbbings and quiver-ings of the earth that are unfelt by our unaided senses, and draws conclusions to serve the needs of men.

It is Professor Milne to whom London editors despatch hurrying reporters when news comes from Japan of another earthquake calamity, and he usually corrects their information—as in June, 1896, when Shide was besieged by newspaper men.

"This earthquake happened on the 17th," said they, "and the whole east-

Gifu, Japan, after the earthquake of 1891.

This and the three pictures following are from Japanese photographs reproduced In "The Great Earthquake in Japan, 1891," by John Milne and W. K. Burton.

ern coast of Japan was overwhelmed with tidal waves, and 30,000 lives were lost."

"That last is very probable," answered the Professor, "but the earthquake happened on the 15th, not on the 17th;" and then he gave them the exact hour and minute when the shocks began and ended.

"But our cables put it on the 17th."

"Your cables are mistaken."

And, sure enough, later despatches came with information that the destructive earthquake had occurred on the 15th, within half a minute of the time Professor Milne had specified. There had been some error of transmission in the earlier despatches.

Again, a few months later, the newspapers published cablegrams to the effect that there had been a severe earthquake at Kobe, with great injury to life and property.

Railroad track twisted by the great earthquake in Japan in 1891.

"That is not true," said Professor Milne. "There may have been a slight earthquake at Kobe, but nothing that need cause alarm."

And the mail reports a few weeks later confirmed his reassuring state ment, and showed that the previous sensational despatches had been grossly exaggerated.

Professor Milne is also the man to whose words cable companies lend anx ious ear; for what he says often means thousands of pounds to them. Early in January, 1898, it was officially reported that two West Indian cables had bro ken on December 31, 1897.

"That is very unlikely," said Professor Milne; "but I have a seismogram showing that these cables may have broken at 11.30 A.M. on December 29, 1897." And then he located the break at so many miles off the coast of Haiti.

This sort of thing, which is constantly happening, would look very much like magic if Professor Milne had kept his secrets to himself; but he has given

The work of the great earthquake of 1891 in Neo Valley, Japan.

them freely to all the world, and for a year or more has been making every effort, with the encouragement of the British Association for the Advancement of Science, to have earthquake observatories established at various points on the earth's surface, with instruments similar to his own, so that by comparison of records, fuller knowledge may be had of movements in the earth's crust and changes in the ocean's bed.

And various governments, universities, and learned societies, quick to see the importance of such knowledge, have sent favorable replies, so that now Harvard University, at Cambridge, Massachusetts, has its own earthquake observatory; Yerkes Observatory at Williams' Bay, Wisconsin, is expected to have one shortly; New Zealand is putting up two; South Africa has one, at Cape Town; Toronto, Canada, has one; India has three; Japan has one; Mauritius has one; South America has one, in Argentina; Beirout, in Syria, is in correspondence for one, and so also is Siberia.

In short, there seems to be little doubt that within a few months no fewer than twenty of these seismic stations will be in operation in different parts of the globe, all equipped with the Milne instruments, and all in regular communication with the head, or central, station at Shide. It is taken as certain that a comparison of records from all these earthquake observatories will make it impossible for an important seismic disturbance to occur anywhere, whether on land or under the sea, without its precise location being immediately known, as well as all essential facts regarding it. And when it is borne in mind that at present seventy-five per cent. of the whole number of earthquakes occur in the bed of the ocean, the value of such statistics to cable companies (and what country is not interested in the proper working of ocean cables?) is at once apparent.

Twice, for instance, it has happened in Australia (in 1880 and 1888) that the whole island has been thrown into excitement and alarm, the reserves called out, and other measures taken, because the sudden breaking of cable connections with the outside world has led to the belief that military operations against the country were preparing by some foreign power. A Milne pendulum at Sydney or Adelaide would have made it plain in a moment that the whole trouble was due to a submarine earthquake occurring at such a time and such a place. As it was, Australia had to wait in a fever of suspense (in one case there was a delay of nineteen days) until steamers arriving brought assurances that neither Russia nor any other possibly unfriendly power had begun hostilities by tearing up the cables.

PROFESSOR MILNE'S LIFE AND EXPERIMENTS IN JAPAN.

Before explaining the workings of these wonderful seismic instruments which are to do the world such famous service, I will tell how it happened that Professor Milne became a student of earthquakes; for, unlike any poetry, seismology is not a career that men are born to. In the Professor's own words: "It was Japan that did it, and that famous cable-laying American, Cyrus Field." Mr. Field heard of Milne back in the seventies, when the young Lancashireman had just finished his studies at King's College, London, and the School of Mines, and was casting about him for such work as the world might have for him to do. He had no more idea then of becoming an earthquake specialist in Japan than he had of hunting pigs in Borneo. Yet he lived to do both. Mr. Field had inquired at the School of Mines for a bright, competent young man who could go out to Newfoundland in the service of the cable company and locate some coal fields for them. Milne was selected, and told to report at a

Effect of the great earthquake of 1891 on the Nagaragawa Railway Bridge, Japan.

certain office in the city.

"I am glad to see you, sir," said the millionaire, when Milne was shown in. "We want to know if you can sail for Newfoundland on Tuesday next?" This in the most matter-of-fact tone and with scarcely prelude.

Milne was fairly at a loss for words; he was barely twenty-one, and had but small experience in business matters. Finally he managed to ask about compensation.

"There will be no trouble on that point," said Mr. Field; "you can leave a memorandum on Monday of what you want for your services; I dare say it will be satisfactory. The point is now, can you sail on Tuesday?"

That was Friday, and Milne pointed out that the shops closed early on Saturdays, and on Sunday he could get nothing, so he was uncertain whether he could be ready in time.

At this, Mr. Field leaned forward on his desk, and said, with a look half serious, half quizzical, that Milne never forgot: "My young friend, I suppose you have read that the world was made in six days. Now do you mean to tell me that, if this whole world was made in six days, you can't get together the few things you need in four?"

Milne was silent a moment, and then said: "I'll be ready, sir, on Tuesday." And so he sailed for Newfoundland—and what he did there is a separate chapter. But it was all to his credit, for soon came an offer from the Japanese Government, intent upon getting the best brains in Europe to assist in the nation's development, inviting Milne to join its service, at a handsome salary, in the department of mines and public works.

So it came about, twenty-five years ago, that this young Englishman took up his abode in Tokio, and in due course turned his attention to earthquakes. This happens quite naturally when one finds oneself in a country where there are two or three earthquakes a day on an average, counting small and large, throughout the year, and where in many instances a single one of these earthquakes has been a more serious matter to Japan in loss of life, and almost as serious a matter in resulting expenditure, as her recent war with China.

Under such circumstances, it was not difficult for a keenly interested and scientifically-trained European to develop into an earthquake enthusiast; and Milne was soon putting forth seismic theories with the best of them, and trying experiments with rough-and-ready seismoscopes and seismometers, which were sometimes rows of pins propped up in a certain way, so that in falling they would give indications as to wave direction, or sometimes bits of string with weights at the end designed to act as recording pendulums; or, again, gravestones tumbled over on their sides in the hope that by their slide or shifting they would show the line and intensity of the earthquake movement.

He produced plans of earthquake-proof houses: houses with roof-timbers running down to the floor sills, which was equivalent, practically, to having the roof rest on the ground. He also showed the Japanese engineers how to build bridges with parabolic piers, so that at any horizontal section they offer equal resistance to effects of momentums applied at the base.

And, as the value of his conclusions became apparent through actual tests, the Japanese Government, properly grateful, established a chair of seismology at the university, and picked Milne out as the one best qualified to fill it; which meant that here was a young man, fresh from a country where there are no earthquakes, officially appointed to teach people who had lived among earthquakes all their lives what earthquakes are, and what measures should be taken against them—in short, the whole business of seismology.

Then began an interesting set of experiments, carried on for years by

Professor Milne, with artificial earthquakes, which he could turn on at will by touching an electric button. Dynamite was used here, buried in the ground, and exploded when the seismographer was ready. Sometimes he would set off five or six of these little earthquakes at one time, and take the records with a like number of seismographs placed at different distances, and connected electrically, so as to show the rate of wave transmission. Once the Professor, in his eagerness to watch the seismograph at the very moment of shock, placed himself within twenty feet of a mine, his position being barricaded by earthworks, with an old door over the top to keep off falling stones.

When all was ready, he waved his hand to an assistant who stood at some distance ready to send the current. Bang! went the dynamite like a broadside of heavy cannon, and the Professor had scarcely fixed his eyes upon the moving smoked-glass disk with the little recording fingers on it, when about a ton of earth came smashing down upon the door, flattening out man and instrument, and bringing that experiment to an untimely end.

On another occasion, at the command of the emperor, a seismic exhibition was organized in the palace yard, where a number of miniature towns and villages had been laid out neatly for the purpose of being blown

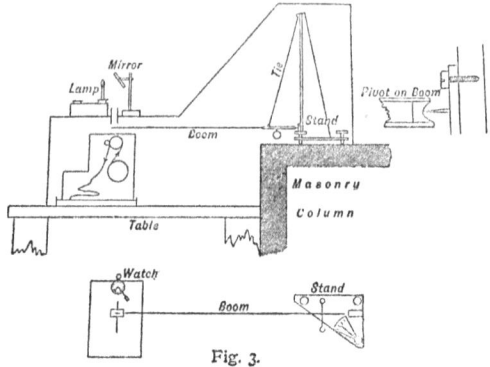

Fig. 3.

Diagram showing vertical and horizontal sections of the more sensitive of Professor Milne's two pendulums, or seismographs.

Professor Milne's sensitive pendulum, or seismograph, as it appears enclosed in its protecting box.

The sensitive pendulum, or seismograph, as it appears with its protecting box removed.

up and shaken down when his majesty should touch the button. Everything went off perfectly, and the courtiers were delighted. For twenty years Professor Milne carried on his experiments, and success seldom failed him. Then he returned to England.

Coming now to Professor Milne's instruments and their work at Shide, I will repeat what may have been already understood, that they are designed to record movements in the earth coming from distant, not near-by, centers of disturbance; they would be of no more service for an earthquake within a hundred miles of them than a telescope would be at the theatre. The seismographs used all over Japan record earthquakes that can be felt; the Milne horizontal pendulums record earthquake waves that cannot be felt. After years given to the practical side of seismology, Professor Milne is now studying its theoretical side, although, as has been seen, much practical good is resulting from his investigations.

THE EARTHQUAKE OBSERVATORY.

My first view of the instruments was at night. Professor Milne walked beside me, carrying a lantern, and his Japanese assistant, Shinobo Hirota, who is nicknamed "Snow" on the Isle of Wight, went ahead to open the doors of the strong-walled little houses where the pendulums were guarded. There are two of these pendulums, both constructed on the same principle, but the one more sensitive than the other. "Snow" showed us the sensitive one first; and when I saw it, I saw only a little lamp burning on a red box with steps to it. The box covered the pendulum. The whole place suggested some silent altar with undying flame. I could hear a clock ticking inside the box.

"What is the lamp for?" I asked.

"To photograph the end of the boom," said the Professor. "It lets a point of light down through that slit. When the earth moves, the boom swings."

"Oh," said I. "And what is the clock for?"

"The clock works the machinery. I'll explain it in the morning, and show you how 'Snow' develops the seismograms."

"Snow" looked pleased, and led the way to the other little house. Here we found a pendulum that was not covered up. It rested on a heavy column of masonry, and one end of it pressed a tiny silver needle against a vertical band of smoked paper that moved slowly between two rollers. There was a clock ticking here also, but no little lamp.

"This," said the Professor, " is an everyday pendulum, to let us know if anything is happening. If there is, then we look at the other pendulum for fuller details. The other one is not so easy to get at. Just glance along that paper band and you can see if there has been an earthquake anywhere in the last twenty-four hours. No, there has been nothing; the line is straight; see—that long white line—the needle makes it as the band turns."

"Suppose there had been an earthquake?"

"I'll show you what would have happened. Come around here; that's right. Now press against the column, not hard, just with your hand. There it goes. See?"

It was like pressing against a chimney, but the boom of the pendulum responded instantly, and the needle swerved out on the paper and then back again, marking a narrow loop.

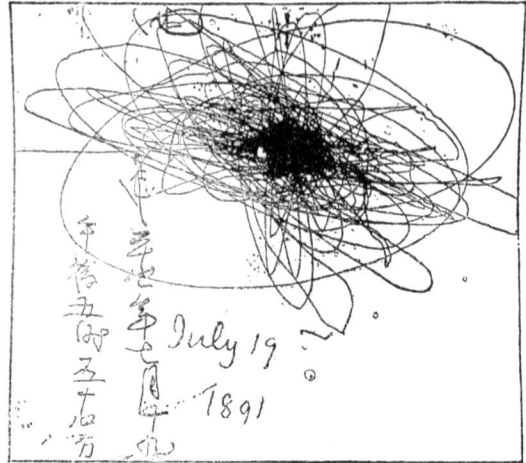

Record made on a stationary surface by the vibrations of the Japanese earthquake of July 19, 1891.

Showing the complicated character of the motion (common to most earthquakes), and also the course of a point at the center of disturbance.

"You tipped the column and altered its level just as an earthquake wave from Japan or Borneo would have done. That is the whole purpose of these instruments, to indicate slight changes of level. They are sensitive to a difference in level of one inch in ten miles. That's not a very steep grade, is it?"

And then he went on to tell how a pair of these pendulums, placed on two buildings at opposite sides of a city thoroughfare, would show that the buildings literally lean toward each other during the heavy traffic period of the day, dragged over from their level by the load of vehicles and people pressing down upon the pavement.

"All these tons of weight make the earth's surface contract between the two rows of buildings, and that tips them together just as you tipped this column. You see the earth is so elastic that a comparatively small impetus will set it vibrating. Why, even two hills tip together when there is a heavy load of moisture in a valley between them. And then when the moisture evaporates in a hot sun, they tip away from each other. These pendulums show that."

I listened in wonder, and presently we went back into the house, which is a real corner of Japan, with a Japanese servant salaaming about and bringing in pleasant things to drink, and the Professor's wife, a Japanese lady, doing the honors with all the grace of her own country.

And the Professor gave some amusing reminiscences of their troubles in getting the instruments properly set up. To begin with, there were imperceptible air currents that would set the booms swinging in a most perplexing way; and when these were disposed of, there came the ghost of Charles I. out of its dungeon and blew the little lamp out, being displeased, so the neighbors declared, at their invading old Carisbrooke Castle (as they did) with such unholy contrivances. After much vain conjecture over this lamp incident, "Snow" finally discovered that it was the doing of a small beetle, which had managed to drop down the tiny glass chimney from the castle ceiling and get himself burned to ashes before extinguishing the flame.

Next there appeared upon the scene—or rather made himself felt—a little gray "money-spinner," that managed to hide inside the red box and would come out nights for experiments of his own. This little spider knew nothing about earthquakes, but took the greatest interest in the swinging of the boom, and soon began to join in the game himself. He would catch the end of the boom with his feelers and tug it over to one side as far as ever he could. Then he would anchor himself there and hold on like grim death until the boom slipped away. Then he would run after it, and tug it over to the other side, and hold it there until his strength failed again. And so he would keep on for an hour or two until quite exhausted, enjoying the fun immensely, and never dreaming that he was manufacturing wonderful seismograms to upset the scientific world, since they seemed to indicate shocking earthquake disasters in all directions.

Such yarns as these the Professor spun for me that evening in his charming Japanese-English home, and he showed me photographs of earthquakes in Japan, taken by himself and his friend Professor Burton, and pictures of volcanoes blowing their heads off, and he told me of exciting adventures crossing Iceland with a remarkable man named Watts, who would jump across yawning chasms just to see if he could do it. Finally, we went to bed.

The next day gave me a better understanding of the instruments, and a good idea of the regular routine of work in an earthquake observatory. I followed "Snow" through his ordinary round in the little houses, saw him wind the clocks that keep the record bands moving, glance through the slit in the red box to make sure that the boom was swinging free, fill the lamp, see that the watch which marks the hours on the band was right to the second, mix some fresh developer for the films, and then, for my especial benefit, draw the red window, and develop the accumulation of four days, a strip about fifteen feet long, which might have on it a record of earthquake horrors, or might

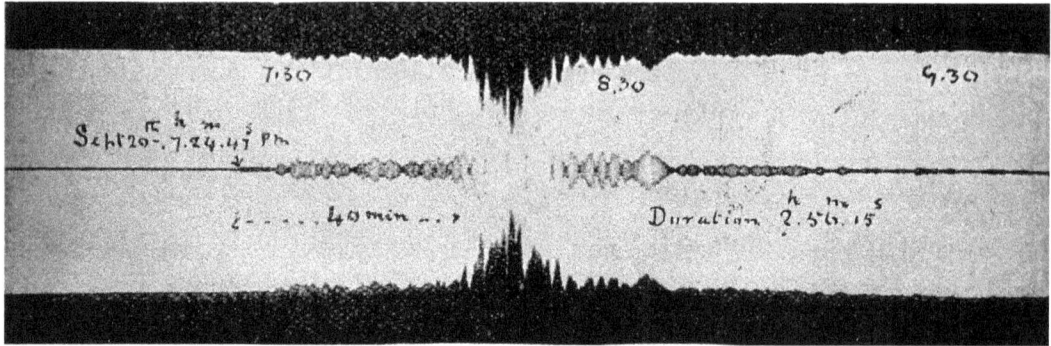

Seismogram of a Borneo earthquake that occurred September 20, 1897.

have nothing. You can never tell until the end of the week, when in the ordinary course a batch of seven days' films is developed. In this case there was nothing, only a straight line down the length of the band. The earth had been behaving itself. But they showed me other films from other weeks that indicated a very different state of things.

As for the instruments, I saw that they are simple enough in principle, though most admirable in perfection of adjustment and delicacy of working. Beautiful devices they are, to do for our sense of level, if I may so express it, what the microscope does for our eyesight. A horizontal pendulum, or boom, poised against a knife edge at the base of a mast, that is the essential feature. A wire stay from the masthead supports the far end of the boom, and a weight hung from it keeps everything taut.

Then two backscrews allow either leg of the supporting tripod to be raised or lowered by the thickness of a spider's web, and even so small a change of level as that disturbs the end of the boom. And that makes the point of light move on the band of paper, and that movement is photographed, so that the record shows a slight loop. As nothing is allowed to disturb the boom, once the pendulum is adjusted, it follows that if the record band shows loops and curves instead of a straight line, it is because the earth's surface has moved underneath the supporting column and changed its level.

As a matter of fact, the earth's surface moves very frequently with tremors like a creature of life, and with long heavings caused by distant seismic disturbances. And for each of these movements the pendulums give an individual record with characteristic waverings and loops on the band, and queer ups and downs that mean nothing to the inexperienced eye, but everything to the seismologist. When "Snow" brings in news of something on the band, there is excitement in that quiet house at Shide as among waiting tiger hunters at a crashing in the jungle.

In each of these records the time is marked in hours along the edge of the band, this being done automatically by hourly passage of the long hand of the watch over the slit in the red box, that shuts off the light for an instant and makes a line on the photographic film.

When a man finds himself in the midst of such an unfamiliar subject as earthquake shocks that cannot be felt, he naturally asks questions, and I asked a great many during my stay at Shide. For instance:

"Does the ground really move, Professor, when these waves come from the other side of the earth?"

"Undoubtedly; it rises and falls just as the ocean does. You see, the earth's crust is very elastic; it is constantly quivering and pulsatory, I might almost say breathing."

"How much does the ground rise and fall with one of these waves?"

"Oh, about three inches."

"What! the solid earth comes up three inches right under us and then goes down three inches?"

"Certainly, it does that very frequently."

"But why don't we see it or feel it?"

"Because it moves so slowly and evenly; fifteen seconds, perhaps, for the lift, and as many more for the descent. And then the waves are so long— several miles between two crests—that everything about us rises and falls together; half of all London heaves up and settles down with a single breathing."

"And how long does it take these waves to travel around the earth, say from Japan?"

"They don't travel around the earth—they travel through the earth; that is one of the most important discoveries we have made. If they were transmitted in the earth's crust around the circumference we should get two records for every earthquake—one coming the shortest way round, the other coming the longest way; for, of course, these wave-movements would be propagated in both directions. Waves through

Pieces of a submarine cable picked up in the Gulf of Mexico in 1888. The kinks are caused by seismic disturbances, and they show how much distortion a cable can suffer and still remain in good electrical condition, as this was found to be.

the air, for instance, from volcanic explosions, always come to us both ways around the earth, the one being recorded after the other. Do you see that?"

"Yes."

"Well, we never get two records of earthquakes, we only get one; so we conclude from this and other reasons, that the transmission is straight through the earth. Do you understand?"

"You mean that these waves come to us along the chord of the arc instead of along the arc itself," I ventured, recalling my geometry.

"Exactly, and now I come to the most important thing: we find that all these waves from distant earthquakes reach Shide in practically the same number of minutes, no matter where the earthquakes occur. They come from Japan in sixteen minutes, from South America in sixteen minutes, from Java in sixteen minutes, and so on as far as our data extend. When all the stations are working, we shall be able to verify this conclusion; but it certainly looks already as if the period of wave transmission through the earth was uniform."

"I don't see, Professor, if all these different earthquake waves get here in the same time, how you can tell one from the other, or know that this one started in South America and that one in South Africa, and so on?"

"I may say in a general way," he replied, "that we know them by their signatures, just as you know the handwriting of your friends; that is, an earthquake wave which has traveled 3,000 miles makes a different record in the instruments from one that has traveled 5,000 miles, and that again a different record from one that has traveled 7,000 miles, and so on. Each one writes its name in its own way, as you have seen on the bands. It's a fine thing, isn't it, to have the earth's crust harnessed up so that it is forced to mark down for us on paper a diagram of its own movements!"

"Are these differences in the wave signatures due to differences in the distance traveled?"

"Exactly. See here, I can make it plain to you in a moment."

He took pencil and paper again, and dashed off an earthquake wave like this:

"There you have the signature of an earthquake wave which has traveled only a short distance, say 3,000 kilometers; but here is the signature of the very same wave after traveling, say, 9,000 kilometers.

"You see the difference at a glance; the second seismogram (that is what we call these records) is very much more stretched out than the first, and a seismogram taken at 12,000 kilome-

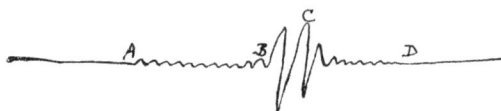

ters from the start would be more stretched out still. This is because the waves of transmission grow longer and longer, and slower and slower, the further they spread from the source of disturbance. In both figures, the point A, where the straight line begins to waver, marks the beginning of the earthquake; the rippling line AB shows the preliminary tremors which always precede the heavy shocks, marked C; and D shows the dying away of the earthquake in tremors similar to AB.

"Now it is chiefly in the preliminary tremors (we call them the P.T.'s) that the various earthquakes reveal their identity. The slower waves come, the longer it takes to record them, and the more stretched out they become in the seismograms. And by carefully noting these differences, especially those in time, we get our information. Suppose we have an earthquake in Japan. If you were there in person you would feel the preliminary tremors very fast, five or ten in a second, and their whole duration before the heavy shocks would not exceed ten or twenty seconds. But these preliminary tremors, transmitted to the Isle of Wight, would keep the pendulums swinging from thirty to thirty-two minutes before the heavy shocks, and each vibration would occupy five seconds.

"There would be similar differences in the duration of the heavy vibrations; in Japan they would come at the rate of about one a second, here at the rate of about one in twenty or forty seconds. It is the time, then, occupied by the preliminary tremors that tells us the distance of the earthquake. Earthquakes in Borneo, for instance, give P.T.'s occupying about forty-one minutes, in Japan about half an hour, in the earthquake region east of Newfoundland about eight minutes, in the disturbed region of the West Indies about nineteen or twenty minutes, and so on."

"Then, really, the information you get from the seismogram is simply that an earthquake has occurred somewhere at a certain distance from the instrument?"

"Yes; but that is quite sufficient to locate the earthquake with absolute precision, since the other stations working with us have similar information. So many miles from Shide, and so many miles from Batavia, and so many miles from Argentina, and we must, with the help of a pair of compasses on the map, fix the place beyond question. And that is why it is desirable to have as many observatories as possible in different parts of the earth. Who can say, for instance, what great sums might be saved cable companies if they knew the precise boundaries of danger re-gions in the ocean's bed?"

"Are such regions well marked?"

"So well marked that a blind man could pick them out by running his fingers over a map of the ocean's bottom made in relief. Wherever he found sudden slopes going down from hundreds to thousands of fathoms, he could say with confidence, 'There is one.' We know in a general way some of these dangerous regions—there is one off the west coast of South America from Ecuador down; there is one in the mid-Atlantic, about the equator, between twenty degrees and forty degrees west longitude; there is

Professor John Milne.

From a photograph by S. Suzuki, Kudanzaka, Tokio.

one at the Grecian end of the Mediterranean; one in the Bay of Bengal; and one bordering the Alps; there is the famous 'Tuscarora Deep,' from the Phillip-pine Islands down to Java; and there is the North Atlantic region, about 300 miles east of Newfoundland. In the 'Tuscarora Deep' the slope increases 1,000 fathoms in twenty-five miles, until it reaches a depth of 4,000 fathoms.

"There have been submarine earthquakes here, like that of June 15, 1896, that have shaken the earth from pole to pole; and more than once different cables from Java have been broken simultaneously, as in 1890, when the three cables to Australia snapped in a moment. And the great majority of breaks in the North Atlantic cables have occurred at the place just indicated, where there are two slopes, one from 708 to 2,400 fathoms in a distance of six-ty miles, and the other from 275 to 1,946 fathoms within thirty miles. On Oc-tober 4, 1884, three cables, lying about ten miles apart, broke simultaneously at the spot. The significance of such breaks is greater when you bear in mind that cables frequently lie uninjured for many years on the great level plains of the ocean bed, where seismic disturbances are infrequent."

Then the Professor went on to explain in detail how the cables are broken by these submarine earthquakes, the two chief causes being landslides, where enormous masses of earth plunge from a higher to a lower level, and in so do-ing crush down upon the cable, and "faults," that is, subsidences of great are-as, which occur on land as well as at the bottom of the sea, and which in the latter case may drag down imbedded cables with them. Statistics show that fifteen breaks in Atlantic cables between 1884 and 1894 cost the companies about $3,000,000, and it is estimated that if the whole coast line of the world was looped with cables, as may be the case some day, there would be not less than three hundred interruptions annually from seismic disturbances.

It is evident, then, that as the laying of ocean cables increases, it is of the first importance that cable companies be in possession of the best available knowledge as to the more dangerous regions in the ocean's bed and the safer regions. This knowledge can come only through the study of such phenomena as are being investigated now at the earthquake observatories of the world.

About the Authors and Editor

Ray Stannard Baker was a journalist, biographer, and author. He was born in Lansing, Michigan on April 17, 1870, where he also went to college and law school. He began his journalism career at the *Chicago News Record* in 1892, then joined the editorial and writing staff of *McClure's Magazine* in 1898. He also wrote fiction, children's stories, and a series on rural living under the name "David Grayson." In 1906, he, Ida Tarbell, and Lincoln Steffens left McClure's and started *The American Magazine*. Baker wrote *Following the Color Line* in 1908, the first examination of the racial divide in the United States by a well-known journalist. He also developed a friendship with Woodrow Wilson, eventually writing 15 books about Wilson that included an eight-volume biography. The last two volumes earned Baker the Pulitzer Prize for biography in 1940. Baker died July 12, 1946. Several of his books as "David Grayson" can be found at Project Gutenberg: www.gutenberg.org/ebooks/author/929

Sir Robert Ball served as Royal Astronomer of Ireland from 1874 while at the University of Dublin, Dunsink Observatory. He was born July 1, 1840 and died November 25, 1913. In 1892, he became the Lowdean Professor of Astronomy and Geometry at Cambridge University. He wrote popular articles and books and gave lectures about astronomy and space, thanks to his simple style. He was the son of naturalist Robert Ball. You can find some of his works at Project Gutenberg: www.gutenberg.org/ebooks/author/866

Henry J. W. Dam, also writing as Henry Dam and H.J.W. Dam, was a journalist and playwright born on April 27, 1856. He died April 26, 1906. In addition to his work with *McClure's*, Dam worked for the London bureau of *The New York Times* and then the London edition of the *New York Herald*. After it was discontinued, he moved into drama. Dam wrote the plays "The Shop Girl," "The Coquette,"and "The Silver Shell," about nihilism, dynamite, and Russia. He was married to actress Dorothy Dorr.

Henry Drummond was a Scottish evangelist and writer, a protégé of D.L. Moody, and a scholar of natural sciences. Born August 17, 1851, Drummond studied at the University of Edinburgh and joined the Free Church of Scotland. He lectured on natural science at the Free Church College, then traveled to Africa and Australia. His Lowell Lecture in Boston inspired McClure to have Drummond write for the magazine. He died in 1897, and became known for his treatises *The Ascent of Man* and *The Monkey That Would Not Kill*.

E.J. Edwards was the byline of investigative journalist Elisha Jay Edwards, best known as the reporter who broke the suppressed story of President Grover Cleveland's secret cancer surgery in 1893. Edwards was born in 1847, graduated from Yale University in 1870 and law school in 1873. He worked for the *New York Sun*, the *New York Evening Sun*, wrote a column for *The Philadelphia Press*, *Chicago Inter Ocean*, and *The Cincinnati Enquirer*, and published article in *McClure's Magazine*. He was a friend of author and poet Stephen Crane, who he also wrote about. After Edwards wrote the story about Cleveland's surgery, the White House began a fairly successful smear campaign to disgrace Edwards, which was only mitigated because one of Cleveland's doctors came forward with the truth. Edwards died in 1924.

Samuel Pierpont Langley, writing as S.P. Langley, was an aviation pioneer, inventor, astronomer, and the third Secretary of the Smithsonian Institution. After college, he eventually became director of the Allegheny Observatory, where he improved the equipment and began distributing standard time to the railroads. Using precise astronomical observations, Langley created the Allegheny Time System, including time zones. He brought in substantial money selling the time service to railroads and cities, which helped finance observatory upgrades and his research on solar phenomena. As part of that work, Langley invented the bolometer, an instrument for measuring infrared radiation and precise changes in temperature. The "langley" is a unit of solar radiation named after him. Langley then directed his intellectual effort to powered flight. He built heavier-than-air craft with rubber-band tension and showed that a craft could have sufficient lift yet fly a distance. Working with Charles Manly, Langley led the development of a powerful engine for the aerodrome, but the Wright brothers flew first and Langley abandoned the project after several crashes. Claims by the Smithsonian that the Langley aerodrome, flown by Glenn Curtiss in 1914, was the "the first man-carrying aeroplane in the history of the world capable of sustained free flight" led to a long feud with Orville Wright. Langley died February 27, 1906. In addition to the "langley" unit, he has an award, a mountain, a U.S. Air Force Base, a NASA research center, and several ships named after him.

Cleveland Moffett was a journalist, playwright, and translator. Born April 26, 1863, he graduated from Yale College. Moffett worked at the *New York Herald* from 1887-1892 as a foreign correspondent, then as foreign news editor at the *New York Recorder*. He returned to the *Herald* as Sunday editor in 1908. Meanwhile he wrote for *McClure's Magazine* along with other publications. In addition to his science articles, Moffett's true crime adventures of the Pinkerton Detective Agency ran in *McClure's*. Moffett died October 14, 1926 in Paris. He also produced several plays, story collections, and novels, some of which can be found at Project Gutenberg:
www.gutenberg.org/ebooks/author/2850

Ida Minerva Tarbell edited and wrote for *McClure's Magazine* starting in 1893. She was born November 5, 1857 to Franklin Tarbell, a small oil producer and refiner in Pennsylvania. His business failed due to collusion between the railroads and larger oil interests like the Standard Oil Company. This partly inspired Ida Tarbell to write a scathing exposé of John D. Rockefeller and Standard Oil's practices which ran originally in *McClure's Magazine*. She became known as one of the "muckrakers," journalists (many from *McClure's*) who dug deep and reported on corruption in government and industry. Tarbell also increased the subscribers for *McClure's* significantly with her long and detailed series on Abraham Lincoln and Napoleon Bonaparte. A successful author and suffragist for much of her life, she died in 1944 and was inducted into the National Woman's Hall of Fame in 2000. A couple of her works can be found at Project Gutenberg: www.gutenberg.org/ebooks/author/4022

Editor

Larry D. Clark is managing editor of *Washington State Magazine*, the award-winning research and alumni publication at Washington State University. Previously he worked for the Washington State Legislature, nonprofits, and a magazine in Japan. Clark has degrees in journalism and Asian studies from Washington State University and the University of Oregon. His interest in the science writing of *McClure's* came from research into fiction from magazines of the late nineteenth and early twentieth centuries.

Notes

Illustrations

All but one of the illustrations are original artwork from the *McClure's* articles. The exception is the portrait of Alexander Graham Bell, which is from the Library of Congress:

> Title: Alexander Graham Bell, three-quarter length portrait, standing, facing right, by window
> Date Created/Published: [ca. 1902]
> Reproduction Number: LC-USZ62-92864 (b&w film copy neg.)
> Call Number: LOT 11533-B-3-21 [item] [P&P]
> Notes: Gilbert H. Grosvenor Collection of Photographs of the Alexander Graham Bell Family.

Not all of the illustrations from the articles are in this collection. *McClure's Magazine* editors had a strong fascination with images of the inside of houses, and I really didn't think people needed to see Maxim's kitchen or Röntgen's closet. I chose not to include those illustrations and a handful of other incidental or repetitive images.

Sources

Daniels, George H., editor, *Nineteenth-Century American Science: A Reappraisal.* Northwestern University Press. 1972.

Johnson, Eric Michael, "The Uses of the Past: Why Science Writers Should Care About the History of Science—And Why Scientists Should Too." *Scientific American* blog, retrieved January 17, 2012. http://blogs.scientificamerican.com/primate-diaries/2012/01/17/uses-of-the-past/

Knight, David, *The Nature of Science: The History of Science in Western Culture since 1600.* Andre Deutsch, London, 1976

McClure's Magazine. Vols. 1-13. June 1893-October 1899.

McClure, S.S., *My Autobiography.* Frederick A. Stokes Company, New York. 1914.

Sharlin, Harold I., *The Convergent Century: The Unification of Science in the Nineteenth Century.* Abelard-Schuman, London, 1966

Wilson, Harold S., *McClure's Magazine and the Muckrakers.* Princeton University Press. 1970.

Type

The article titles and subtitles are Linux Libertine Regular, a public domain font by Philip H. Poll and the Linux Libertine Open Fonts project. More information can found at linuxlibertine.org.

The cover slab face is Kelson Sans Bold. Kelson is a remake of the original Kelson type family, designed by Bruno Mello from São Paulo, Brazil. You can see more of Bruno Mello's work here:.
http://www.myfonts.com/person/Bruno_Mello/
Kelson Sans Bold is a Creative Commons licensed font.

List of Illustrations

www.ingramcontent.com/pod-product-compliance
Lightning Source LLC
Chambersburg PA
CBHW080720220326
41520CB00056B/7156